Gooney Bird Driver

© 2019 Jon A. Maguire. All rights reserved. No part of this book may be scanned, copied, uploaded or reproduced in any form or by any means, photographically, electronically or mechanically, without written permission from the copyright holder.

ISBN: 978-1-943492-67-1 (Hardback)
ISBN: 978-1-943492-68-8 (Soft Cover)

Cover photo likely taken by Joe Stevens.
Colorized by Jakob Lagerwij, Piece Of Jake Concept Art, Netherlands.

The events described in this book are true and factual as recalled and told to the author by Joe Maguire to the best of his recollection. All stories are based on his memories of actual events, locales, people and conversations. All conversations are based on Joe Maguire's recollections and are not intended to represent word-for-word transcripts but are presented in a way that evokes the essence of what was being said.

Elm Grove Publishing | San Antonio, Texas | www.elmgrovepublishing.com
Elm Grove Publishing is a legally registered trade name of Panache Communication Arts, Inc.

Gooney Bird Driver

The stories of WW2 C-47 pilot Joe D. Maguire and the combat missions that led to his honors and awards decades later

JON A. MAGUIRE

Dedicated to my mother, Jeanan, who always held us together.

Contents

Preface 7

Introduction 8

1. Before the War 11

2. WAR 15

3. Into Combat in the ETO 26

4. The 27th Air Transport Group 33

5. Grove 52

6. Le Bourget 58

7. Villacoublay 83

8. Clothing and Equipment 107

9. Almost Fifty Years Later... 111

Appendix 1: Aircraft Credit Sorties 118

Appendix 2: Airfield Locations 121

Acknowledgments 128

The Author 129

Gooney Bird Driver

Preface

THERE HAVE BEEN COUNTLESS books published on WW2 pilots of fighters, bombers and troop carrier aircraft. One group that has largely been overlooked is that of combat cargo and medical evacuation flyers. These men and women pioneered, on a large scale, what became "medevac." As the Allies were driving across Europe after D-Day, the job of supplying the needs of gasoline, combat cargo and evacuating wounded was an overwhelming task. Many of these young pilots were pulled from Troop Carrier Command because they were highly trained in C-47 operations. They were placed into the 27th and 31st Air Transport Groups under the 302nd Transport Wing, Air Service Command, USSTAF. They flew unarmed to makeshift landing zones on the front lines, unloaded gasoline or cargo, and loaded wounded that had to be evacuated regardless of weather conditions. They were often under fire during these critical operations. These groups also performed countless other non-combat transportation of VIPs, POWs, emergency operations and even a secret operation in Sweden. They played a vital role in the Battle of The Bulge, flying in replacements to the front lines. They also had many interesting assignments including transporting the Glenn Miller band. My dad was one of these pilots. I grew up hearing his stories and those of his buddies in the 27th Air Transport Group. I became their historian in the 1990s and wrote their unit history. My dad, Joe Maguire, passed in 2016 and I realized that his stories needed to be preserved as a vital part of WW2 history. I had made a number of recordings of him and I also had a number of his stories ingrained in my mind. I have told his stories in his words and placed historical context around them to form his military biography. Joe D. Maguire was awarded the DFC, Air Medal with six oak leaf clusters, and the French Légion d'Honneur. He was credited with 35 combat missions and had roughly 1000 theater hours in the ETO.

Note: The Douglas C-47 aircraft was affectionately known as "Gooney Bird" to those who flew her.

Introduction

Joe Maguire was a depression era kid from Oklahoma City. Like most kids of that time period, he grew up poor. Joe's parents, H.G. and Mary, were divorced, which was very unusual for that era. He was a guy who loved passionately and had a fun, silly side to him. He had a deep faith in Christ, which directed his life, yet he had an "Irish Temper" that could turn on a dime. He was a Baptist Deacon and a man of honor. He entered into a business partnership that ultimately failed because of an alcoholic partner. Many would have declared bankruptcy. Joe took on two jobs, one from 8 to 5 and a second from 6 to 9 until he paid it back. Mom worked too and she always had food on the table for us. Joe had a lifelong passion for golf and he wished I did, but I hate the game. I did share his joy when he made a hole-in-one. He gave the certificates and ball to his grandson, Sean. He loved his family and always did the best he could. He was a C-47 Pilot in WW2. Joe Maguire was my dad. I grew up in the 1950's and 60's. It seemed to me that everyone's dad was a WW2 vet, so I did not think it was unusual at all that he was a pilot. On my block, Tom and Elsa Lout were both veterans. Elsa had been a nurse in North Africa, so her house was our local emergency room. Whenever a kid got hurt, they went to see Elsa, so she could patch them up. Whatever happened, it did not compare to what she dealt with in wartime. She treated GI's (Americans) and Tommies (British soldiers). She also treated German POWs. (Dad would call them "Krauts", as was common slang during the war). Her husband Tom was a military hospital administrator and later became a pharmacist. Tom would bring our prescriptions home from his pharmacy. When their son, Randy and I played army, he wore a real Afrika Korps M43 cap that a POW had given his mother. Ken Manner, who lived on the corner, was an 8th Air Force Navigator. Mr. Rea was a B-17 Pilot. Smokey Berthalot was a Marine. He had a Japanese Samurai sword hanging over his fireplace, which made a huge impression on me at the time and probably contributed to me becoming a collector of Militaria. Maurice "Mac" McNeill, who lived across the street, was an Army Air

Force radio operator. I never heard Mac talk about the war. My sister married his son Steve. I know I am forgetting a few, but that was my neighborhood. War stories were commonplace. My dad would tell me a story every night at bedtime, before we prayed together. I heard those stories for years and it got to the point that I could tell them almost as well as he could. I would look through his wartime photographs and he would tell me about the guys in the pictures. I felt like I knew them all.

Sometime around 1990, mom called me and said that Joe Stevens had called my dad! Joe was a pilot in the 321st Transport Squadron, 27th Air Transport Group and my dad's best friend from the war years. They had flown a C-47 across the North Atlantic together in August of 1944. Joe was in town for a 27th Air Transport Group reunion. Before that time, we did not know there was a veteran's organization for the group. That began my journey of attending reunions with mom and dad every year and participating in events until 2011, when the group finally disbanded. I ultimately became the group's official historian and wrote their unit history, *Gooney Birds & Ferry Tales, The 27th Air Transport Group in WW2,* Schiffer Publishing, 1998.

From the time I was old enough to read, I was frustrated by the fact that there was virtually no information about the 27th in print. There were many volumes written about bomb groups, fighter groups, even troop carrier, but nothing on the combat cargo/medical evacuation groups in the ETO (European Theater of Operations). I corresponded with well-known historian, Roger Freeman back in the 1990s. Roger sent me a note, thanking me for identifying the markings on the C-47s of the 27th. He remembered seeing them as a child in England, but had never identified the unit. That was my main reason for writing the history – I wanted them to be remembered.

As dad got older, his memories faded, but I still knew the stories and could prompt him. When mom and dad went shopping, he always wore his Distinguished Flying Cross Society, C-47 or USSTAF (United States Strategic Air Forces) hat. Mom would lose track of him, only to find him telling war stories to a stranger. On Christmas Eve, he would always tell us the story of the Battle of the Bulge and the part he played, flying in replacement troops on Christmas Eve, 1944. Joe Maguire passed away on January 24, 2016. This past Christmas, I thought about how much I missed hearing that story. As a kid in the neighborhood in the 60s, the stories were plentiful. Now, they are gone… they must be remembered.

I will put some historical context around those stories to create his military biography. I have several recordings of him, though many of his tales are stored in my brain somewhere. To the best of my ability, I will tell them just as he did. His words are presented in italics. Some of the punctuation may be a bit strange, but where I had recordings, I wrote the words as he spoke them, so as not to lose the character of his speech patterns. Off We Go!

1. Before the War

JOE WAS BORN IN OKLAHOMA CITY, Oklahoma on October 5, 1923 to Mary Ellen and Harvey G. Maguire. He had two older brothers, Chester (Chet) and Harlan (H.L.). Chet was 14 years older and H.L. was 11 years older than Joe. Chet became a tanker in WW2 under Gen. Patton's command, but we never heard much about his service. H. L. was seriously injured in a car accident and was not able to serve in the military, but he worked at Will Rogers Army Air Field in Oklahoma City during the war years. He was a plumber by trade and his service was valuable to the efforts at home. As a child, Joe experienced a number of hardships, close calls and illnesses. His parents divorced, which was almost unheard of at that time. At age 7, he had his appendix removed on the kitchen table. He carried a scar for life from that procedure that extended halfway across his abdomen. One time Joe decided to look into an empty gas can, using a match. Of course the fumes ignited and he was lucky that he was not blinded, or worse. He also had a bout with Scarlet Fever, which at that time required isolation. They would not even let his mother in to see him. With tears streaming down her cheeks, she could only look up at the window of his room, from the ground below. It was a different time. People were tough. H.G. ran a plumbing business with several Model T Ford trucks. As an infant, Joe was sitting in the driveway playing. One of the men did not see him and backed a truck over him. Fortunately, the Model T axles were high enough that it just bumped his head. Joe was lucky he even lived long enough to make it to WW2!

Economic times were tough as well. In 1929, the Great Depression hit and people were out of work. As a plumber, H.G. could usually put food on the table. If people could not pay their bills, they could at least trade for services. The stress of the depression finally got to H.G. and he suffered what was then referred to as a nervous breakdown. Mrs. Maguire did what she could. She took in ironing, sewing, and ran a boarding house. Joe did not have a bed of his own until he joined the Army. He grew up sleeping on a folding cot. Many of the American soldiers in WW2 came from similar backgrounds as a result of the depression.

From the time he was very small, Joe wanted to fly! He dreamed of flying a SPAD or SE5a. He cut out newspaper clipping of airplanes and saved them. When the Curtis P6E Hawk went into service, it became his favorite airplane. The Smilin' Jack aviation comic strip started in 1933. He looked forward to reading that in the funny papers! Joe also built model airplane kits when he could save enough money to buy one. At that time period a kit consisted of blocks of wood with templates. The builder would sand the wood using the templates until they matched. Then they would be assembled with wood glue and painted. Joe's mother would help him dry the glues and paint by putting them in a low temperature oven for a short time. He used to tell of a time when he had seven models in various stages of completion. He had seen his mother use the oven so he decided to do it himself. He carefully placed the models on a tray and placed it in the oven. All was well until he went outside to play. As you may have guessed, they were all destroyed from being left in the oven. That was a very hard lesson for a small boy to learn and it probably contributed to him becoming a very meticulous person. Many would call him a perfectionist. He set a very high standard for himself and his family. Sometimes one that was unattainable.

In 1939, Joe was able to attend the World's Fair at Treasure Island in San Francisco. They drove from Oklahoma City to California. At the fair, he saw his first significant, modern, military aircraft: The B-17 Flying Fortress prototype. That only increased his desire to become a pilot. Today we cannot imagine what it must have been like to see that mighty silver modern aircraft. To him, it must have looked like something out of Buck Rogers (a fictional space character of the period).

At age 12, Joe met Jeanan Joyce, who would become the love of his life. They rode on his bicycle together that day. They later attended high school at Central High in Oklahoma City and were married at Sedalia Army Air Field on June 12, 1944, shortly before he shipped out to England.

Joe loved Big Band music and he loved to sing. Until his last days, he knew the words and music to every tune by Glenn Miller, Tommy Dorsey, Benny Goodman and many other bands. He was thrilled that he got to fly the Miller AAF Band some during the war. As he was in hospice care, toward the end of his life, he listened to his bands and sang along with his favorite tunes. It was his request that Big Band music be played at his funeral. His request was, of course, carried out. He also asked me to sing "Danny Boy" (Londonderry Air) at his funeral. He was proud of his Irish heritage. I knew I couldn't get through it live, so my good friend Josh Eddington at Crossings Community Church helped me record it and I gave it to him before he died. He approved.

B-17 Flying Fortress prototype at the World's Fair in San Francisco, 1939.

Joe's best pal growing up was his cousin Ralph Jenks. Ralph's mom, Celia, was Joe's mother's sister. They did everything together. He and Ralph enlisted, shipped out, completed part of training together and ultimately both ended up as pilots in the 27th Air Transport Group, but in different squadrons. That was a pretty unusual chain of events. Their other pal who enlisted with them, Dolph Farrand, ended up flying B-25s in the CBI (China Burma India) with the 490th Bomb Squadron, "Burma Bridge Busters."

Joe worked at Oklahoma City Air Depot, which later became Tinker AFB, before he went into the Cadet program. He would later fly a C-47 across the North Atlantic that was built there…

Ralph Jenks, Joe Maguire and Dolph Farrand while they were still together at Shepherd Army Air Field in Basic Training.

2. WAR

Joe usually started his stories with *December 7, 1941, Pearl Harbor…* Many American veterans talked about that day, as it was a seminal moment for all of them. Like many of his contemporaries, Joe spoke of gathering with family by the radio to hear President Roosevelt's address describing, "A day that will live in infamy."

There was no question for most young American men that they would enlist and get into the fight as soon as possible. Joe was a junior in high school at the outbreak of war. He wanted to be sure he got his chance to fly, so he decided to enlist after graduation. When he went to the recruiting office in downtown Oklahoma City, it did not go as planned:

> *I went down to the recruiting offices and went straight to the Marines, because I wanted to fly a Corsair. When I got there and began the procedure the Marine Sergeant said, "I'm sorry son, but you are too small." I was devastated and started to leave, when he called me back in. He said, "I believe if you will go eat bananas and drink as much water as possible, you can make weight for the Army Air Force, which is just down the hallway." So I did that and barely made it in. I had a 27" waist and wore a size 36 short jacket at that time. I enlisted about November of '42. I graduated from Central High School, Class of '42. I was called to active duty in January of '43… I guess it was. Until we got on flying status we did not wear the cadet uniform. We started wearing that once we were in ground school for flying. We went into Wichita Falls, Texas from our homes and went through basic military training. Then we were sent to a college training detachment at Texas A&M, where it was mostly trying to brush up guys on math and physics. (Referring to his photo album) This is Leland Pope, a kid from high school who was killed after he got his commission. Someone*

told me he was flying a P-40 in training. They were still using them in training before going to planes like the P-51. These are just shots of the training planes. (Another photo) *This is a kid about my size named Johnny Long. I have never seen him again since we got separated. We didn't go to the same Advanced Training. They kind of broke up people alphabetically. There must have been a break between the L's and M's so he went someplace else. It also may have depended on what you indicated you wanted to fly. I never did fully understand. These are more Parks Air College shots and a bunch of the guys I trained with. For the most part, I never saw them again after that. You remember these barracks when we went up there and you got to see them. I was really fortunate to get to go through Parks.*

Joe was a small guy and made these comments about the challenges of flying:

> *There are cases where someone may have been shorter than me but not many because I believe 5'6" was the breaking point. I was 5'6" at the time. I don't know if the military still has standard flight cushions or not, but due to my size, when I would go fly those little airplanes, I would check out four cushions – three to sit on and one to my back because I could not reach the controls on some of the airplanes. Some of the bigger guys would laugh at us smaller guys but there were a lot of guys that used them. It was funny when I got into the big aircraft that were more sophisticated. They had more seat adjustments so I didn't need as many cushions.*

Two years later, Joe's separation papers indicate that he was 5'6" and weighed 134 lbs.

Opposite page:
Top left: Johnny Long and Joe Maguire in Primary Flight Training at Parks Air College in East St. Louis.

Top right: Joe Maguire in dress Cadet uniform at Parks Air College.

Bottom two pictures: In Joe's words: *These were taken at my mother's house down on 1627 NE 12th Street in Oklahoma City. I was home on leave. Larry was born in '43, so that has to be late '43 or early '44.* His mother, Mary; brother, H.L.; sister-in-law, Marie and nephew, Larry, are in the photos also.

Joe with a Fairchild PT-26. This or the PT-19 would have been the first type of aircraft flown in Primary, before moving to Basic Training.

The proud Air Cadet points himself out on snapshots sent home to the family.

One of the other cadets in Joe's class poses by a PT-19.

Fairchild PT-19 with shark-mouth nose art at Parks Air College. The 19 and 26 were the same aircraft, except the 26 had an enclosed cockpit with a canopy.

I entered the military in January of 1943 and I went through basic at Sheppard Field in Wichita Falls, Texas, where I learned to be a soldier, then to a College Training Detachment (CTD) at Texas A&M - from there, to San Antonio at Randolph Field. My Primary was at Parks Air College in East St. Louis, Basic was in Independence, Kansas, and I graduated at Frederick Army Air Base in Oklahoma, Class of 44D. I went to Sedalia, Missouri, for combat training. After that final phase of training, everybody gets to have leave and go home. Then I returned and got sent to Baer Field in Ft. Wayne, Indiana, with 1st Troop Carrier Command. I wanted to fly the P-38 (twin engine fighter) *so I requested multi-engine training. That is how I ended up at Frederick in my home state of Oklahoma. When I completed training the 38 was being phased out and that's how I ended up in the C-47. I was disappointed at the time but it turned out to be a good thing. That Gooney is a good old airplane! They are still using them!*

Left: Joe with a Cessna AT-17 Bobcat. This was the main twin-engine trainer at Frederick, although Joe did have some time in an AT-9 "Jeep". He was proud of having flown the Jeep. He always added, *"That airplane had a higher wing-loading than a B-26!"*

Below: *These photos were taken at my brother's house in Oklahoma City, which was across the street from what is now Northwest Classen High School, where my kids and nephews graduated from in the '60s and '70s. I had graduated from Frederick AAF, Class of 44D. I gave my graduation wings to Jeanan, which she is wearing in the picture on the left.*

Left: Jeanan shows off Joe's graduation wings.

Below left: *Warrensburg, Missouri. Jean and I had a little old one-room place and that was it. We got married in June. I graduated in April at Frederick (44D) and then was transferred up to Sedalia. Mom came up there and we got married in the base chapel there. These pictures were made right before we got married I believe. I was not old enough to get married so I had to have the permission of my commanding officer.*

Below right: With his buddy Joe Stevens in Warrensburg.

Jeanan and Joe at the time of their wedding.

Above: *Although we look like a bunch of messes; we were all commissioned officers there. That is after we got out of Frederick and were learning the C-47 and Troop Carrier Command. We learned all the Troop Carrier tactics and worked with paratroopers. These pictures were out on a bivouac, like you were out in a combat zone. This was near Sedalia, Missouri.*

Below: *It would not be much longer until we were in the ETO...*

We completed our transition training and were ready to enter service in Troop Carrier Command. During that time I flew large formations and dropped paratroopers. In Troop Carrier we would fly huge formations. The prop wash would get so bad it was like riding mush. You always wished you were up front. I towed gliders and did "snatches" where we would pick up a glider from a dead stop, by flying low and hooking a cable with a tail hook, attached to the glider, which was stretched between two poles. It felt like that old Gooney was going to stand still in midair! Then as the nylon rope recoiled and the glider caught up, the airplane would speed up. I also flew gliders. I flew the snatch in a C-47 and in a Waco CG-4. They said it took six seconds for the glider to go from a dead stop to flying. There was a rumor floating around that if you had so many hours, you would be considered a qualified glider pilot, so I made sure I did not get very many hours! I wanted no part of that. My friend, Joe Stevens, was a glider pilot, before he became a pilot.

3. Into Combat in the ETO
(European Theater of Operations)

AFTER SEDALIA, WE WENT *home on leave and then had to report to Baer Field, Indiana. Baer was home to 1st Troop Carrier Command. At Baer we were prepared to enter the war. We had been issued all lightweight Pacific Theater gear and I was scared they were sending me to the jungle. I hate all kinds of bugs and mosquitoes! I also hate humidity. I suppose it was all part of the plan to keep us in the dark. We were then sent to Dow Field in Bangor, Maine. One night they collected the Pacific gear and issued winter weight flight clothing for the ETO. Shortly after that we were given our orders. They handed us an envelope marked SECRET and we were told not to open it until one hour after departure from the United States. It was all cloak and dagger stuff.*

Note: In February of 1942, Dow Army Airfield was transferred to Air Service Command (ASC). It was located near the Air Transport Command (ATC) North Atlantic air ferry route to the United Kingdom. Its primary mission was servicing long-range combat aircraft before they flew across the North Atlantic to RAF Prestwick, Scotland. In March of 1944, Dow AAF was transferred to Air Transport Command's North Atlantic Wing.)

One interesting side note – During the war, Hollywood actors of that period were very supportive of the war effort. The military used a lot of celebrities in training films and other areas to support the troops. The guy that conducted our overseas briefing was actor, Hume Cronyn. He was married to Jessica Tandy, who was also a Hollywood actress. Major League Baseball Hall of Famer, Enos Slaughter of the St. Louis Cardinals, used to hit us fly balls when we were in basic training.

25 August 1944 - We proceeded as ordered with my buddy, Joe Stevens as first pilot and I flew co-pilot at that time. That navigator showed up with a big basket of sandwiches and a thermos of coffee. He had made the trip many times.

For the trip across the North Atlantic via the Northern Route, each crew was assigned an ATC navigator. The navigator on that trip was Charles I. Dreben. They never saw him again after that trip. Normally C-47 crews did not have a navigator.

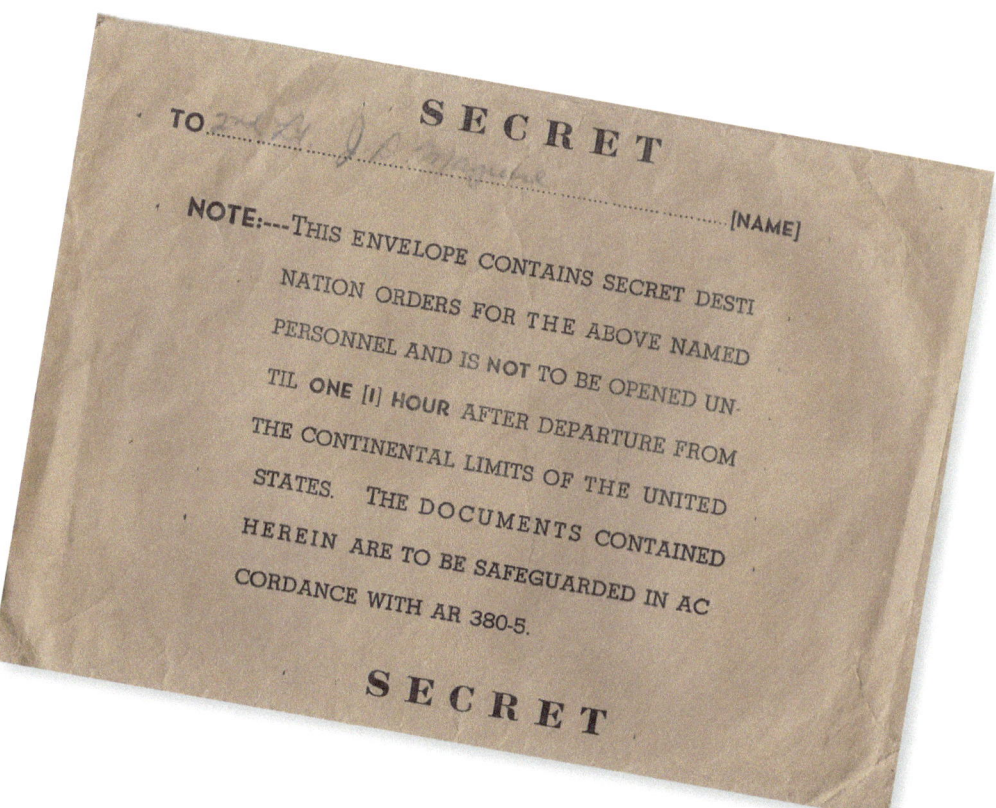

SECRET envelope containing orders for the crew given to Joe Maguire, the co-pilot of the aircraft.

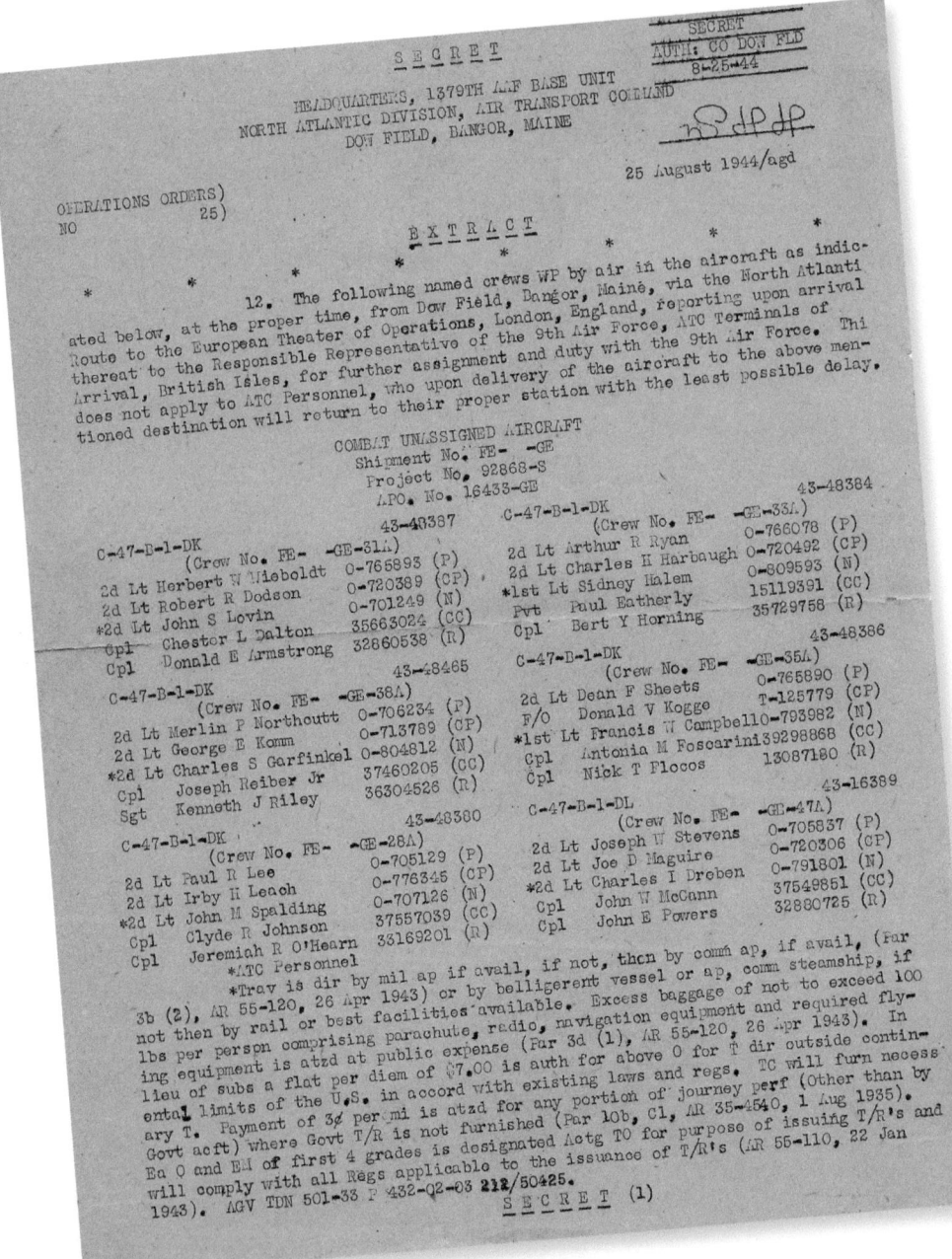

SECRET orders instructing the crew to fly "via the North Atlantic Route to The European Theater of Operations, London, England, reporting upon arrival to the Responsible Representative of the 9th Air Force, ATC Terminals of Arrival, British Isles, for further assignment and duty with the 9th Air Force. This does not apply to ATC Personnel, who upon delivery of the aircraft to the above mentioned destination will return to their proper station with the least possible delay."

SECRET

Operations Orders #25, Dow Fld, Bangor, Maine 25 August 1944.   P2

This is a PERMANENT change of station.

In lieu of subs a flat per diem of $7.00 is auth for T and for periods of temporary duty enroute to final destination, when necessary for O & F/O, in accordance with existing law and regulations. Payment of mileage is not auth. Such times as the individual is billeted and subs as outlined in WD memo W 35-2-42, 30 Sept 42, his per diem shall cease.

A flat per diem of $7.00 is auth for EM for T and for periods of temporary duty enroute to final destination, in accordance with existing law and regulations if T is performed by air. For T by rail and for periods of delay enroute to final destination, monetary alws in lieu of rations and quarters is prescribed in accordance with AR 35-4520.

From time of departure from the continental United States until arrival at permanent overseas station, payment of per diem is auth for a maximum of forty-five (45) days.

Preparation for Overseas Movement AAF Replacement Combat Crews (Short Title POCR)-dtd 1 April 1944, constitutes an integral part of this order.

TDN. 501-31 P 431-02-03 212/50425.

AUTH: Ltr fr CG AAF to CG ATC, Sub: "Asgmt and Reasgmt of Mil Pers of the A F", 12/1/42 and 1st Ind fr CG ATC to CG MAW, 12/10/42 and AR 55-120 par 3b (2) 1943 and Unnumbered TWX AFOCR, Wash 25, DC, 2200Z, 22 Aug 44, and Under auth Ltrs: CG ATC to CO, Fer Div, 5 Nov 1942, Sub: "Plan of Organization and Operations for Foreign Deliveries of Aircraft, ATC 300, 16 May 1944; and 1st Ind CG Fer Div 12 June 1944: (and any other auth such as T.X).

* * *

By order of Colonel JENSEN:

ROBERT L SIMONS
Major, Air Corps
Senior Aircraft Operations Officer

OFFICIAL:

WILLIAM H. HULBERT
Captain, Air Corps
Ass't Operations Officer

DISTRIBUTION: CG NAD-2; CG AAF-4; EDATC-2; CO Sta 16-1; PO-1; Ass't Opns O-1; Stat Sec-1; Ea indiv conc-6; Opns O Sec File-13.

SECRET
-2-

Note: The Navigators were Air Transport Command. The rest of the crews were Regular Army Air Force.

> *John McCann was our crew chief and John Powers was radio operator. That North Atlantic Route was a long, tough flight. From Baer Field, Indiana – Dow Field, Maine - Goose Bay, Labrador - Greenland - Reykjavik, Iceland – RAF Nutts Corner. The C-47 we were flying was brand new. We did the final testing on it at Ft Wayne, to accept it into service for the military. Not just us, but every crew was assigned a plane to take to Europe that hadn't been completely tested out yet. Our plane had been built here (Oklahoma City) at Tinker Field and had 18 hours on it.*

Upon landing, the crew did not receive much of a welcome to the ETO. Instead they were introduced to the black market and the realities of a war zone.

> *Some colonel and a sergeant came up to the plane in a jeep and took the survival gear off of the airplane - the hand crank radio, shot guns, fishing equipment - all the survival equipment. They took it under the guise that it was their official duty to do so. We were just a couple of brand new second lieutenants, so we didn't question a bird colonel. I think Joe Stevens had signed for all that stuff and later on they tried to hold us accountable for it. Somehow we got out of having to pay for it. I'm sure it was sold on the black market.*

This part of Joe's story was always a bit cloudy in his memory. He was a 21-year-old kid (in his own words) and things were happening very quickly. Another event happened, which drastically changed his direction in how he would serve. After having been trained in all of Troop Carrier tactics and operations and having orders that directed him to report to 9th AF headquarters in London, on 5 September 1944, a number of crews were pulled from this group and reassigned to the 27th Air Transport Group. This likely occurred at RAF Heston. The 27th ATG originated in the 8th Air Force and ultimately became part of the 302nd Transport Wing, Air Service Command, United States Strategic Air Forces (USSTAF). This was shortly after the D-Day invasion and the drive across Europe was getting underway. The need for combat cargo, medical evacuation, gasoline and all types of logistical support was critical. There was a huge need for well-trained C-47 pilots outside of Troop Carrier Command. Joe always described this period as being sort of a blur. He remembers arriving in the ETO and then a long journey on a train. The next clear memory he had was being at RAF Heston, in London, for a very brief period, and then arriving and being stationed at Grove Field, near Wantage, England. The places and dates where Joe was stationed are approximately as follows:

>28 August 1944, arrival at RAF Nutts Corner, Northern Ireland
>31 August 1944, Heston Aerodrome, London, England
>6 September 1944, Grove Field, near Wantage, England
>2 November 1944, Le Bourget, Paris, France
>21 February 1945, Villacoublay, Paris, France
>8 May 1945, Germany surrenders
>3 June 1945, Camp Detroit assembly area near Laon, France
>10 August 1945, Boarded train for Marseille, France
>10 August 1945, Dijon France, Railway station –
>>learned Japan had surrendered
>
>11 August 1945, Camp Calais, Near Marseille, France
>23 August 1945, Boarded USS General Morton
>2 September, 1945, Arrived at Newport News, VA,
>>Camp Patrick Henry

Lt. Joe D. Maguire would fly 35 Aircraft Credit Sorties, commonly referred to as "combat sorties" or "missions," with the 321st Transport Squadron, 27th Air Transport Group, 302nd Transport Wing, USSTAF (United States Strategic Air Forces) for which he received the Distinguished Flying Cross and Air Medal with 6 Oak Leaf Clusters. (He did not receive the medals until 1994). While he was credited with those combat sorties, (133:10 combat hours) he also flew 633:50 additional theater hours, which were important missions, just not considered combat. These regular Army transport units were extremely vital to the war effort, but they received very little recognition at the time, or in the history books. Joe received battle stars for Central Europe, Northern France, Ardennes and the Rhineland campaigns.

A C-47 of the 27th ATG taking off from a Marston Mat temporary airfield. (USAAF)

4. The 27th Air Transport Group

THE 27TH AIR TRANSPORT GROUP and the 31st Air Transport Group were unique in the ETO and operated differently from other US Army Air Force units, so it is important to Joe's story to understand a brief history of the group.

As the war progressed, it became apparent there was not a cohesive structure in place to serve the transportation, medical evacuation and combat cargo needs across the theater. The logistics involved in supplying Patton's drive across Europe was nothing short of miraculous. The 27th ATG started life in the 8th Air Force to handle ferry and transport services. It ultimately ended up under the 302nd Transport Wing, along with the 31st ATG, under USSTAF, serving the supply and transport needs of the European Theater of Operations (ETO) and beyond. The 31st ATG was formed under the 9th Air Force. The 27th ATG was made up of the 86th, 87th, 320th, and 321st Transport Squadrons, and the 310th, 311th, 312th, 325th and the 2920th (Provisional) Ferrying Squadrons. By late 1944 the 311th and 312th were functioning as C-47 transport squadrons. There were also two Service Squadrons, the 519th and the 520th.

27th ATG pilots and crewmen in temporary quarters, June 1944, near the front lines.

Radio operator Bill White at 27th ATG Headquarters, RAF Hendon.

The group participated in many significant operations. One of the most notable was the Battle of the Bulge, where the 302nd Transport Wing received a commendation from Gen. Spaatz. The 27th played a major role in support of Patton's drive across Europe, and they performed a Top Secret operation in neutral Sweden, supporting the Norwegian underground. The Swedish missions were conducted under the command of noted Norwegian-American aviator, Colonel Bernt Balchen. The biggest danger on these missions was not always the enemy-it was the weather! Temperatures were often around 40 below zero. There were also some interesting encounters on these neutral bases, as in some situations the Luftwaffe was also flying in and out of the same airfield! Maguire did not participate in the Swedish missions but several of his friends did.

Opposite page (bottom) and top two pictures this page: C-47 Operations in Sweden. Makeshift heaters were used to keep the oil from freezing.

Below: Norwegian troops boarding a C-47 in Sweden.

Armed Luftwaffe soldiers at the gate to the airfield at Bodo, Norway on VE Day.

The history of the unit is a difficult story to tell. Unlike fighter and bomb groups, which were stationed at one air base, squadrons from the 27th and 31st ATGs were operating out of multiple bases across the theater. I have met several men who did not know what squadron they were assigned to at a given time and many were on detached service, which made it even more difficult. It was a unit that literally went wherever it was needed at the time. In the wartime 302nd Transport Wing unit history, the 27th Air Transport Group was called "The Workhorse of The Wing." Joe Maguire always said, "We were just flying truck drivers," but that was an understatement...

There is much confusion about Air Transport Command and the 27th and 31st Air Transport Groups. (Sometimes ATC was "lovingly" referred to as "Allergic To Combat.") The 27th and the 31st ATGs were never part of ATC. There were some ATC C-47s reassigned to the 27th, as were many former Troop Carrier ships. As a result, photos show C-47s in operation with a hodgepodge of markings visible, until they had a chance to repaint them. These groups were under Air Service Command, United States Strategic Air Forces and were the only units of their kind in the ETO.

Another difference in the operation of these C-47 units from those of Troop Carrier or Bomber crews was that they did not fly with the same crew or aircraft every day. In Joe's words:

> We didn't fly the same airplane everyday. You might fly one number one day and another one the next. You took what they gave you. We trained as a crew and were a crew for a while but when we got worked into that unit (27th ATG) everybody was part of a pool.

You waited until the operations officer made the assignments. I can see where that worked out to develop camaraderie. It let everybody know everybody. Everybody flew with everybody and it let everyone get to know each other. Crew chiefs were mixed around and got to meet different guys. Nobody felt strange like that other bunch. (Speaking of the pilots that were already in the unit when Joe's group arrived in theater.) *The only way you got to know them was eventually flying with some of them. Of course, they were all the old salts. Every one of us had to make flights with them until we learned the ways over there – the British system. We had a squadron check pilot, operations officer and all that. You had to fly with the squadron check pilot. I flew with him when I got that "Rainbow Card" instrument card* (SCS 51 ILS). *I got checked out and passed that. Most everybody did. For a while, every new pilot flew nothing but co-pilot. Then after while you would trade off so everyone could get first pilot time. We also flew a lot of assignments that were just one or two ships – not large formations, although we did some large ops as well. Once we qualified as first pilots, we would switch seats on round trips so we could both get flying time as first pilots.*

Joe qualified as first pilot in mid-September of 1944.

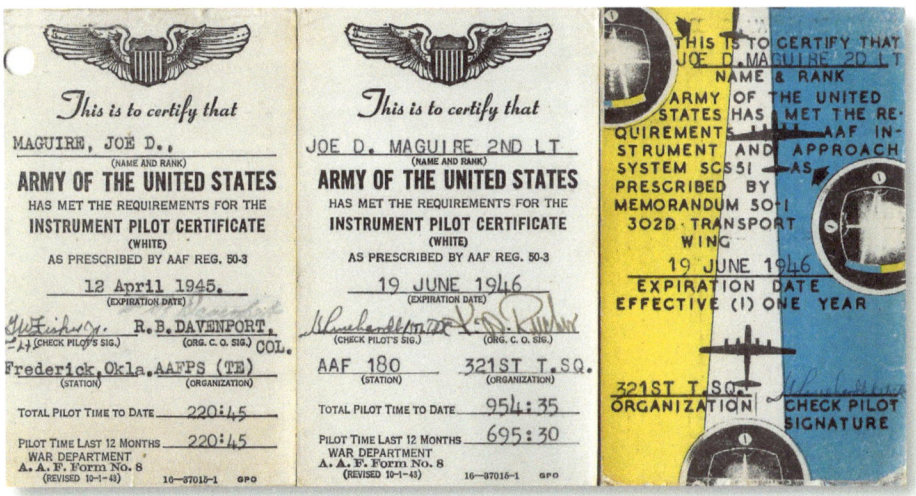

Instrument Pilot Certificates. The card on the right is the one Joe called the "Rainbow Card." It is an unusual card as it is specific to the 302nd Transport Wing. He was always very proud of it. It states that the holder of the card has met the requirements of AAF Instrument and Approach System SCS 51. This was the American Radio Guidance system and it was the basis of ILS (Instrument Landing System). The card features the image of the SCS 51 instrument. Joe said, *"With that card you could over-ride the tower if the base was closed due to weather."*

Sheet from Joe Maguire's flight log illustrating the practice of pilots switching between first pilot and co-pilot on missions, so they could both get first pilot flying time.

For our group, flying Goonies was like driving trucks. We were pretty busy all the time. We were really busy when other units — fighter and bomber, were moving from England to the continent. We were also busy when those battles for France were going on – picking up wounded with nurses and medical technicians. Most of the time we would fly them back to England. It was very busy flying gasoline to points where tanks and vehicles needed gas to operate. We would fly right up to the front lines. We would fly for everybody in the theater. The kind of unit we were, we had to be available for whatever needed to be done. I was just a simple little airplane driver doing what I was told.

Above: Troops awaiting transportation. The C-47 in the background is from the 31st ATG. The white triangle on the tail was the group marking of the 31st. The 27th ATG used a white circle.

Right: Brig. Gen. Howard M. Turner, Commanding General of the 1st Air Division, 8th Air Force, greets combat crewmen on their return to England, 17 Feb. 1945. These air crewmen had been released from an internment camp in Switzerland. The tail markings of the 27th ATG are clearly visible. The yellow and black stripes denoted the 302nd Transport Wing, Air Service Command and the white circle was the group marking of the 27th ATG. (USAAF via Chancellor).

Above: Cargo being delivered to the troops, possibly mail bags.

Below: C-47 of the 86th Transport Squadron making a delivery. Also visible is a UC-64 Norseman which may be from the 320th Transport Squadron and a 36th Troop Carrier Squadron C-46.

Above: A Gooney Bird going to work from a temporary grass airfield.

Below: C-47s in formation. A friend snapped these three pictures while Joe was flying the aircraft.

These pamphlets were made up by the 27th ATG Office of Public Relations and given to troops when appropriate.

This folder, called "Poop Sheet" held confidential information for the group while at Villacoublay. It contained the master list of briefing certificates covering the more frequent routes used by the group. It also contained the latest information regarding radio aids to navigation on the continent and in the UK, call signs, etc. It was important to keep up-to-date information. The Luftwaffe would transmit false signals to throw off navigation.

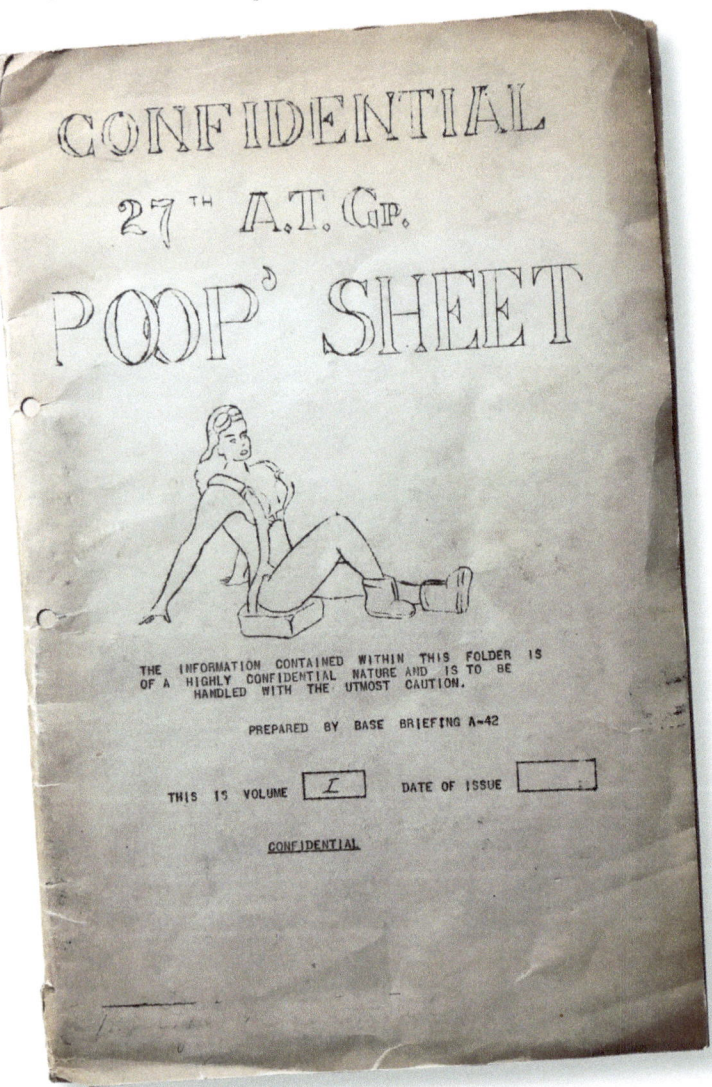

The longest day I ever flew, we took off from Nutts Corner and up around Belfast and from there, clear to Italy – having stops on the way. I'm not real clear if we did that all in one day or maybe we had a night in between. I remember it was about 9:00 at night and we were trying to get a place to stay. I can't remember who was with me. We switched around so much. If you came back to base, you would likely

fly with someone else the next day. The operations officer would post the crews every day. I had 35 combat sorties, but I would fly something about every day. I don't really know what qualified one from the other. I think it had to be a flight into a certain territory or zone of combat. In other words, I made lots of flights from England to Europe but it was in ground that we had total control over. When we flew into areas where we were subject to enemy fire, there was a possibility of being hit by enemy fire, shot down or taken prisoner, that probably constituted a combat sortie.

The 27th also used a different unit code system from C-47s in Troop Carrier. In the 27th, the first letter of the code was the squadron code and the second letter was the individual aircraft. In the 321st squadron, the code was C, so the codes behind the cockpit would read CA, CB, CC, CD, etc. The code for the 86th was T, 311th was S, 312th was P, HQ was W. These were not used until late war. I have not found evidence that the 320th ever used a code. The letters were painted in yellow, sometimes with black shadowing. Only three of the squadrons had squadron patches – the 86th, 310th and 312th.

Left: Don Diehm, right, with a Russian airman he encountered on a joint operation. Don was in the 312th Ferrying Squadron, 27th ATG, which operated like a transport squadron by this stage of the war.

Below: English-made squadron patch of the 312th Ferrying Squadron, featuring a pilot riding a Gooney Bird.

Left: Lt. Nico by the nose of a 312th FS C-47 "Stella" (the name is painted on the bird in the squadron insignia).

Below: The yellow and black-checkered cowlings of the 320th Transport Squadron are visible on this C-47, which has become mired in the mud of a wet grass field.

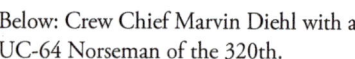

Below: Crew Chief Marvin Diehl with a UC-64 Norseman of the 320th.

Above: C-47 of Headquarters Squadron, "Puppeschon." Note the WA code.

Below: Lt. Maguire in the cockpit of a 321st Transport Squadron C-47. The lemon drop on the cowling was the marking of the 321st.

Top and left: These two C-47s of the 86th Transport Squadron were built in different factories. The Oklahoma City factory applied the olive drab upper surface paint with a wavy line of separation from the gray under surface. The factory in Long Beach used a straight line to separate the colors. This is clearly visible comparing these two photographs

Below: The late war T code of the 86th is visible in this photo.

Above: Close up of the Sad Sack insignia of the 86th Transport Squadron. This insignia is misidentified in a number of publications as the 27th Air Transport Group insignia. It is the insignia of the 86th Transport Squadron of the 27th ATG.

Left: Sad Sack squadron patch of the 86th Transport Squadron. These patches were designed and hand made by Lt. Art Kane who was a pilot.

Left: Lt. Dick Seebers of the 311th squadron. The 311th carried the yellow and black "Tiger Tooth" markings on the cowlings.

Below left: A good example of the tail markings used by the 27th ATG. All aircraft of the 302nd Transport Wing, Air Service Command, carried the yellow-black-yellow stripes. The white dot was on the nose and on the vertical stabilizer. The 27th used the dot and the 31st used a triangle in the same places.

Right: Squadron patch of the 310th Ferrying Squadron.

Left: Commemorative plaque to the 27th ATG at the US Air Force Academy.

Below: 321st Transport Squadron briefing. This photo was likely taken at Grove.

5. Grove

RAF GROVE/AAF-519 WAS AN airfield located approximately 55 miles west-northwest of London near Wantage. It was originally a training field for RAF Bomber Command and it was used for RAF Glider training. The USAAF 9th Air Force began using the field in August of 1943 in preparation for the cross-channel invasion (D-Day). The 31st Air Transport Group was activated at Grove on 28 October 1943. After the invasion of France in June 1944, elements of the 27th ATG were moved from RAF Heston to Grove in order to support the continued invasion. There was a confusing period when the 31st was still under the 9th Air Force and the 27th Air Transport Group was under the 8th Air Force. Finally, the 31st was absorbed under the 302nd Transport Wing of USSTAF. There is virtually no mention of the 27th ATG operating at Grove in many of the history books, nor are they mentioned in the plaque commemorating Grove, but it was here that Joe Maguire began flying operations in support of the Allied drive across Europe.

Joe remembered:

> *When I first got to England, I was assigned to Grove, near Wantage, England. That was one of the main bases of the 27th Air Transport Group. Not only did we service the continent with C-47s, we handled the delivery of aircraft to the theater. Another squadron in the group (310th Ferrying) which my cousin was in, primarily ferried new aircraft and war wearies to and from combat squadrons. My cousin Ralph Jenks has time in about every aircraft in the USAAF inventory. He and I left in the same cadet contingent from Oklahoma City. We were everywhere together until they split us up in San Antonio. We didn't see each other again until I was assigned to Grove. I walked into the mess hall and here is this guy sitting with his neck all wrapped up. He had a lot of trouble with strep throat. He was eating*

soup. I walked in and he recognized me and said, "Well son!" He was younger than me, but he looked older and always called me son. I said, "Ralph!" From then on, we would see each other every now and then. On days off we would try to find each other.

Ralph Jenks and Joe Maguire when they were still Cadets at College Station, Texas. Texas A&M was one of many schools that hosted College Training Detachments (CTD).

When I finally got to the base there in London, the officers club was well established. We went into the officers club and went by the bar to get a drink and then started to mill around a bit. We saw a room that was full of smoke and got close enough that we could hear them talking. As we got closer, we could see it was a poker game and saw the huge stacks of five-pound notes! We visited with some of the guys and I decided I would never play cards with that bunch! They were serious. This outfit had a lot guys in it. Some had flown with the RCAF and RAF and now in the USAAF. One of the guys was named Goodwin. Goody used to say, "I've been promoted five times and I'm still a 2nd Lieutenant." You could fly as a sergeant in their air forces. (RAF) *I wonder what ever happened to Goody.* He said he was going *to marry a Scottish girl and raise chickens. Maybe he did. One of the guys I flew with came back and flew with American Airlines.*

Joe recalled that a short time after arrival in England, he, along with other C-47 pilots, were flown to a Troop Carrier base where they were preparing for a large operation. He told stories of seeing paratroopers practicing their knife-throwing skills around the base as they would spin and hit a wooden target. He distinctly recalled the sound the knife made as it stuck in the wood. He was likely being held in reserve for Market Garden in mid-September of 1944, but he did not fly in that operation.

Left: Joe standing by a Quonset hut at Grove with a message to his wife, Jeanan.

Right: C-47 pilots Lee Powers and Joe Maguire in front of their quarters at Grove.

Left: Joe Maguire wearing his newly custom made Ike jacket. A British tailor cut it down from a dress uniform blouse. These became popular with pilots, as they were easier to fly in. The style was named after Gen. Eisenhower, who frequently wore this style. The British referred to them as battle-dress.

Right: C-47 pilots Bob Bailey and Lee Powers at Grove.

Below: While at Grove, the men lived in this type of Quonset huts. These were common at airfields in the UK.

Just a few days after arriving at Grove, Joe's first aircraft sortie was flown into Villacoublay near Paris. A few months later, the 321st Transport Squadron would be stationed there. This shows just how quickly the Americans were driving across France. The Americans liberated Villacoublay 27 August and Maguire was flying in cargo and evacuating wounded 8 September! The sorties from Grove were all flown across the channel. Joe would talk about these runs fondly.

> *When we would head home over the English Channel, sometimes we would tune in Armed Forces Radio Service. One of my favorite DJs was a guy named Sgt. Monahan. His theme song was Tommy Dorsey's Opus One. I loved hearing that through the headset. Sometimes we would fly over those beautiful White Cliffs of Dover. There was another great wartime song, There'll be Blue Birds Over the White Cliffs of Dover, or I'm Getting Sentimental Over You... I'll be Seeing You... String of Pearls...*

...and then Joe would always sing or whistle a few lines... I'd give anything to hear him sing those songs again.

> *One day we took off on a run and as we were gaining altitude, I noticed a strange shimmer on the wing. I suddenly realized it was gasoline! The crew chief had left the cap off the wing tank spout and gas was sloshing out across the wing. To make it more interesting, we noticed a loose magneto wire arcing in the engine cowling. We executed a long, slow, easy, flat turn and eased the airplane back on the ground. I was sweating like nobody's business. Man I had a few choice words for that crew chief when we were back on the ground!*

Left: Lt. Maguire in the cockpit of a C-47. He sent the photo home to Jeanan. The reverse says, "Note I am in the first pilot seat." That was a big day for a young pilot. Many of them did not live that long.

Right and below: The process of loading a heavy crate into a C-47 using a mobile crane. Note the crane truck has been dubbed "Miss America," visible on the front of the truck.

6. Le Bourget

IN EARLY NOVEMBER OF 19 AND 44 we moved across the channel to Le Bourget in north Paris. It was a big, well-established airfield before the war. It had been occupied by the Germans, so it was heavily bombed. Many of the hangers were just skeletal structures. The roofs were gone or had holes in them. We were billeted in an old school house there in Paris. The Germans had painted large murals on the walls and flying Luftwaffe eagles and swastikas. I wish I had taken pictures.

Photo of Le Bourget taken before the war.

Similar view of Le Bourget taken when we arrived. You can see how much damage there was to the buildings and to the airfield itself. The engineers did a great job of getting it operational in a hurry.

Above: A Gooney in one of the hangers at Le Bourget.

Below: The remains of an Me 109 German fighter in a bombed out hangar

Above: Close-up of the tower at Le Bourget. The damage was extensive.

Above: Destroyed Focke-Wulf 190 fighter.

Below: There is not much left of this Me 410 fighter.

Two aircrew members of the 86th Transport Squadron standing by the Le Bourget sign.

> The Allied drive across France was in full swing and it was a full-time job supporting 9th Army and George Patton's 3rd Army. I remember one time we were at an airfield and someone told us Patton was in the mess hall. We decided not to go in! He did not have a very friendly reputation. We would fly in loads of gas for the tanks and fly out loads of wounded. A lot of time we would land on temporary grass fields or fields with Marston Mat laid on top of it. That was a type of steel planking that the engineers could put down quickly and form a temporary landing surface. Those fields could be hazardous. If a wheel slipped off the edge and sunk into the ground that could cause real problems. Or if the planking was not properly joined or bent, it could blow a tire.

One time I was assigned to fly a general somewhere. It was just another day to us. I was sitting on the tail section of the C-47. I was wearing flight coveralls and GI shoes. It was cold that day so I was wearing a jeep cap (knit stocking cap). *When his driver pulled up to the airplane, I slid off the tail and came to attention. I gave him a salute and he returned it. He said, "Where is the pilot?" I said, "I'm him." The general looked me over, shook his head and said, "Well, let's go son." Once we were in the air, the general asked if he could come into the cockpit. I said, "Sure General." He asked a number of questions about the instruments and he was very interested in how I knew where we were. I explained the instruments and I guess my answers satisfied him. He went back and sat down the rest of the trip. I'm sure we all looked like kids to him and in fact, I guess we were just kids.*

On the edges of Le Bourget, there were some underground German fortifications. We would go out and explore them.

This is Joe by the entrance to the German underground fortifications. *I was clowning around with my .45 and fighting knife. I sent it home to Jeanan. The back says, "Honey don't I look like a rough customer?"*

Another shot of Joe at the entrance to the underground shelter. They must have been flying into a combat zone that day as both Joe Maguire and Joe Stevens are carrying their sidearms.

Joe's best friend from the war years and the First Pilot on the crew from training and the flight across the North Atlantic, Joe Stevens. Joe has his shoulder holster slung around his waist. This shot was taken at the same time they were exploring the German bunkers.

Joe Maguire, Bob Bailey and Lee Powers, all C-47 pilots at Le Bourget. The guys called Bailey "Tadpole." Bob was from Wadley, Alabama. Lee was from Princeton, New Jersey.

Photo taken in front of the school where Joe's unit was billeted. *These guys were all pilots in the 321st Squadron. Most are dressed to go into Paris for the day. I am in the middle wearing a gun-belt, as I was officer of the day. The guy with his hand on my shoulder is Raymond Zych. We called him "Blouse" because he always flew in his dress blouse for some reason. Ray was older than the rest of us and had more street smarts. He would buy diamonds over there and press them into cardboard and mail them home. I always wondered if they made it. That's Art Meeks next to me in the B-10 flight jacket. He must have been flying that day or he would have been dressed like the rest of the guys. I think that's Mayer on the other side of Art. Stevens is second from last. I can't recall the other two guys.*

Another shot of some 321st pilots in their flying gear at Le Bourget. *The first guy is Dick Taylor. He was a Mormon kid from Salt Lake City. Dick was also one of the guys who flew the secret ops to Sweden. I can't remember the second guy. Joe Stevens is in the middle. The tall skinny guy is Messelt. He may have been a RAF guy as he is wearing RAF boots. The last guy is Mayer.*

That's me, Taylor and Bailey.

While flying out of Le Bourget, Joe was involved in several historically significant events. I mentioned earlier that he had a real passion for Big Band music. Perhaps the most well-known band leader of the times was Glenn Miller. Glenn wanted to do his part for the war effort and joined the US Army. Ultimately he formed the Army Air Force Band and played concerts for the troops. One day, Joe Maguire reported to operations and found that he was assigned to fly the Glenn Miller Band to Paris. That was 15 December 1944.

> *I was one of two C-47s assigned to fly the Glenn Miller Band from London to Paris. We flew them to Paris. I remember them talking about how Glenn Miller wasn't on either plane, but he'd be along later. The band went over, half in my plane and half in the other. The story*

was that he was coming over in a Norseman, but we didn't know what he was doing or why he wasn't flying with the rest of us. But he was coming in a Norseman and that's a good little airplane, but that was just one of those things ... I was close to that little bit of history. That was just another assignment. It was just as routine as anything else. You would just go down to the airfield. You would take off and go somewhere and when you got there it could be gasoline, or wounded or whatever. That time, it was the guys that played in the band. They were going over to Paris. They got flown over there and later they got flown down to Marseille. We weren't even aware that Glenn Miller didn't show up. Then it all came out later. It was during the time that he went down. I don't know whether it was that exact day or not but it seems like it was. It seems like that was the initial flight across. The second time we flew them, when we went down to Marseille, was the first time I saw girls in bikinis. It was a put-up deal. The Stars and Stripes newspaper was there. The girls were there to meet the band and they took pictures. Ray McKinley (drummer and co-leader of the band with arranger Jerry Gray after Miller's disappearance) *was on my plane. I get these things mixed up. I can't remember if it was on the Paris or Marseille flight, but it seems like he wanted to know what a Sperry Gyroscope was, so I showed him. I think maybe his wife worked for Sperry?*

(Joe's flight records indicate three landings, 2:45 co-pilot time, 1:15 First Pilot time, in a C-47A on 15 December 1944, the day Miller went missing.)

The pilot of the UC-64 serial number 44-70285, carrying Maj. Glenn Miller and Lt. Col. Norman Baessell, was Flight Officer John R. S. Morgan. He was a small guy nicknamed "Pee Wee." Morgan was a member of the 27th Air Transport Group and had been assigned to several different squadrons. Morgan was also former RAF and had a lot of flying time. Many books and articles have been written about the Miller disappearance so you can draw your own conclusions. Officially they are Missing in Action. They took off on a flight from Twinwood Farm, near Bedford in England and were never seen again.

A casual shot of Maj. Glenn Miller, far left, in the cigarette smoke.

The Glenn Miller Army Air Force Band performing at BAD-2 UK – Warton, USAAF Station 582.

The second historically significant event that Joe Maguire was part of while flying out of Le Bourget was the Battle of the Bulge. Since it occurred over Christmas, he would often tell the story to the family on Christmas Eve. He would also read the Christmas story from the Bible. The Battle of the Bulge was Germany's last major offensive campaign of the war. It took place in the dead of winter from 16 December to 25 January 1945. It is also known as the Ardennes Counteroffensive. It was launched in the Ardennes region, of Belgium, France and Luxembourg. The region had very dense forests, which helped lead to the Germans achieving total surprise. There was also poor aerial reconnaissance due to the weather. The plan was to split the Allied forces and encircle them. The Americans bore the brunt of the attack and took 89,000 casualties. This was the bloodiest battle of WW2 for the Americans and one of the deadliest battles in US history. There were over 1 million men involved in the battle making it the largest engagement ever fought by the US Army. There have been many books written and even a Hollywood movie about this battle.

The 302nd Transport Wing would ultimately receive a commendation from Gen. Carl Spaatz for their actions in the Bulge. Joe Maguire's story would always begin with *It was Christmas Eve 1944...*

A Christmas Card from the 27th Air Transport Group, 1944.

Joe wrote this Christmas note to his brother H. L. and family in 1944.

> The best of Christmas and New Year's wishes Bud, Sis, and Larry from Paris! Christmas 1944.
>
> I would give all, but my Jeanie and life to be with her and all the loved ones at home. Not only at this time, but anytime.
>
> I am with you all always in spirit.
>
> Please remember to utter a little prayer now and then for little Bud.
>
> I love you all very dearly and thank God for such a brother, sister, and nephew as I have.
>
> Love to all
> Joe

The morning of December 24, 1944, I flew to Belgium and returned to our base at Le Bourget in Paris. Everyone in the squadron was looking forward to Christmas Eve celebrations! Rumors were flying that we would get turkey and dressing with all the trimmings, which was a dream come true after all those weeks of K rations, Spam and powdered eggs.

As usual, a truck pulled up to the airplane to take us to our quarters. As I climbed into the truck, I noted a small spotter type airplane with what appeared to be Santa Claus on board. He was dropping sacks of fruit and candy to some local kids who had gathered from the neighborhoods around the airfield. Our headquarters were in a schoolhouse that had been occupied by the Luftwaffe. As the trucks approached the building, here came Santa behind us, still tossing out the gift bags to the kids running beside the truck. I guess it was the same Santa from the airplane. The candy and fruit came from soldiers who had donated it from their rations. Seeing those happy little kids made it all worthwhile.

We went on to our quarters and began to get cleaned up for our Christmas Eve party. Someone had found a tree, which was decorated with various things. It wasn't home, but a lot better than nothing. It was Christmas Eve. Darkness set in and our appetites and expectations were growing, but it was not to be.

We were given strong black coffee and told to get back in our flying gear. Orders came down from higher command that we were to fly down to Istres, near Marseilles in southern France to pick up badly needed tank crew replacements and fly them to advance airfields near the Battle of the Bulge. We flew down the Rhone Valley to the Mediterranean and tried to get some sleep in our airplanes until daylight. At dawn, we could see the trucks coming with hundreds of young soldiers who had just gotten off ships from the States. We were to take them to the front lines to help turn the tide at the Bulge. We flew our load of young soldiers up to a field at Verdun. It was close to the action at the front and it was within range of heavy artillery.

I have been a Christian believer since I was 12 years old at old Kelham Avenue Baptist Church. I recall thinking, what a way to celebrate the birth of the Christ Child, Jesus. I remember wondering why all this war was happening. I was really sad for all those ground troops…

We returned to base, late afternoon on Christmas Day, very tired. I laid on my bunk thanking God for the dependability of the old Gooney Bird and for getting me back in one piece. I thought about the Christ Child again and it reminded me that there is hope! Soon I was sound asleep.

One of my toughest memories from the war happened about a week later. I flew back up around the Bulge area on a medical evacuation flight. My plane was loaded with litter patients and I was walking up the aisle of the plane to the cockpit. A young soldier with his face heavily bandaged reached out and touched me. He said, "Hey Lieutenant, you flew me up here." That memory bothers me to this day.

A series of portraits taken in Paris, near Christmas of 1944. Joe remembered trading cigarettes to French photographer Roger Carlet, who was very well known at the time and had photographed many film stars of the period.

Joe Maguire at Le Bourget, near the time of the Bulge.

A C-47 of the 321st Transport Squadron, 27th ATG, on the field in the snow. Winter 1944.

The 27th did experience tragedy on that Christmas Eve but not on the operation. Several of the men were not on the flight. One of the pilots, Harry Protzeller of the 321st squadron recalled, "Our squadron was quartered in an old schoolhouse about one-half mile from the gate of Le Bourget. On Christmas Eve, a group of nurses invited some of us pilots over to a Christmas party... I don't remember what I drank, but it was too much! Some of our group wanted to continue and decided to get a jeep and go into Paris. I felt horrible and decided to go back to my room and crash out. I was in no condition to continue. Well, they got in the jeep and took off. At this time, the Germans were using English-speaking soldiers to penetrate our lines. As a result the MPs moved the Paris checkpoint out a little further from where it normally was positioned. The jeep load of pilots ran through the checkpoint and the MPs opened fire. Pederson was killed and another man was hit. I guess God looks out for drunks because it is not like me to miss a party unless I am too drunk to go! In this case, it probably saved my life." Harry liked to party until the day he died.

For their actions at the Bulge, the 302nd Transport Wing received a Commendation from Lt. Gen. Carl Spaatz and they were even covered in *The Stars and Stripes* newspaper. The article read:

"Tankers Are Sped To The Bulge by Air At-Peak of Battle.
Hundreds of badly-needed tank mechanics and technicians were flown from rear areas to airfields in the bulge battle zone in less than a day after a rush call had been sent for them at the height of the Nazi counter-offensive, USSTAF disclosed yesterday.

The disclosure was made in a commendation given to the Air Service Command's 302nd Transport Wing by Lt. Gen. Carl Spaatz, USSTAF Chief. The commendation said that 100 planes of the 302nd flew from bases in England and France to a rear-area field on the Continent, picked up the men and sped them up to Third Army on Christmas Day, less than 20 hours after the operation began."

The events of Christmas Eve 1944 were also covered by Ted Malone on his radio broadcast, Westinghouse Presents "Top of the Evening," February 21, 1945.

Malone: Christmas Eve at the 27th Air Transport Group base, like Christmas Eve most places, the fellows were planning on celebrating with a party, but late that afternoon a message came in that changed everything. Col. Feldman was asked if the 27th could possibly furnish planes to fly an army across the Continent to stop the Rundstedt breakthrough. This was like asking a Sabetha, Kansas boy to go to the Worlds Fair. Not that he got excited - I never saw the colonel excited over anything, except maybe Gooney Birds – but this was a challenge. Most of the planes were out on missions. The pilots were scattered. But Col. Feldman knew that the 27th had been asked because other organizations were not prepared. So regardless of difficulties, the assignment was accepted.

Col. Feldman and Covalt walked into Maj. Bennet's office and as casually as though they were delivering a Christmas card, they tossed the order on the operations desk. Bennet read it; looked up, amused. "What's the gag? I don't get it."

Col. Feldman grinned. "Major," he said, "How soon do you figure all our planes will be back at the field here?"

And mimicking his CO's casual mood, looking out at the snow covered fields and threatening skies, "Oh, most of them should be back in a couple of days if the weather breaks."

Feldman stroked his chin thoughtfully. "It won't be easy then, to get them back in a couple of hours regardless of weather."

Bennet's eyes twinkled. "No it won't be easy, especially since the weather is particularly "regardless" tonight."

"It's bad; so is the situation on the ground up around the Ardennes Forrest," the colonel countered.

Suddenly Maj. Bennet became deadly serious. "Is this straight stuff?" Do they really want us to move our army tonight?"

"They're depending on the 27th to do it," Col. Feldman said quietly. "I guess they figured if anyone could, we could and I told them we would. What do you think?"

For answer, Maj. Bennet was already on the phone. He got in touch with each Squadron CO; instructed them to check every plane, pilot and crew available on the field and report back immediately.

Within a quarter of an hour, ground crews were out sweeping snow off the wings of ships; fuel trucks were filling planes with 100-octane gas; pilots were excusing themselves from Christmas Eve dinner parties and climbing into trucks filled with pilots buzzing with questions. "What's up? What happened? What's the idea canceling Christmas Eve?" Surely nobody was expecting them to fly before morning and why rush them in from dinner?

At the same time, planes of the 27th ATG closed for the night at fields all over the Continent, were being opened back up and the motors warmed up for attempted flights back to base, if they could possibly get through.

"What happened?"

"Well the 27th Air Transport Group did it of course,"

Col. Feldman told me. "Bennet got the planes back, checked and refueled. They took off for a field in Southern France where a whole army was waiting. It was the first night flight the pilots had made over France. They went to strange fields they had never landed on, picked up heavy loads, flew them back through the dark to a field near the Rundstedt bulge – and while I can't tell how many planes took the thousands of men, I can say there were enough so that it was generally assumed that there would be some accident, which in an operation of this kind could hardly be avoided.

"But there were no accidents at all. Every ship flew safely through the snow and sleet and delivered its load on schedule and Christmas Morn, a whole new Army stopped the Nazis dead in their tracks.

"That's quite a story," I told the Colonel.

"It's the Gospel truth," he said.

Another significant event in Joe's wartime experience was the loss of a comrade on 18 January 44.

> *All the guys called him "Pork" because he was a little chubby. His real name was Roy Shilling. Pork was always smiling and cutting up - we all loved him. Pork was assigned a routine flight to go pick up some guys who had finished their tour and were heading home. It was just another day.*

The official report reads: 2nd Lt. Roy J. Shilling, pilot, 0-696907, 321 TS. 18 JAN 45 left A-54C (Le Bourget) to go to AAF-343 (Biggin Hill) with 18 GI troops. C-47 took off at 9:23 from A-54C and seven minutes later crashed at Plessis-Gassot, France. All aboard were killed with the exception of four passengers. Reason for crash unknown. Other crew killed were co-pilot, 2nd Lt. George A. Wood, radio operator, Cpl. Alphonse Cogozza and crew chief, Cpl. Harold W. Grubb.

> *One of the guys* (name withheld) *went a little crazy. We found him by himself in his bunk. He had scratched the wall until his fingers were bleeding. We each dealt with grief in our own way. We all loved Pork.*
>
> *Some of the toughest flying we did was flying out wounded soldiers. Often times we would fly in a load of gas or ammunition and*

Left: Joe Maguire and Shilling before a flight together. Right: In Joe's words: *Shilling goofing around on a hoist that was in the courtyard of our squadron billet. He was always cutting up.*

Joe Maguire, Bob Mayer, Roy "Pork" Shilling and Ray Zych. *That winter of 1944-45 was a cold one. We had to cut our own firewood.*

Joe Maguire loved to have fun and make people laugh his whole life. He made up funny songs and invented words. He loved to laugh. This picture captures his personality pretty well. He is obviously hamming it up for the camera. He rarely flew in a helmet with oxygen mask because most of the flights were low-level. Whatever the occasion was, he had the picture taken. The caption reads: Feb. 1945. *I have on my fighter helmet with oxygen mask hanging by my face.* This was in front of the 321st Squadron quarters at Le Bourget.

then here came the ambulances. They would fold out the litters and start loading wounded. Many times, we could not fly very high, especially if we had chest wounds. We were not equipped with oxygen for the patients. I had some real hard days flying, but the most stressful was a day that I picked up some guys from a field hospital in France and we had to take them up to a hospital in Scotland. The flight nurse came into the cockpit before takeoff and told us there were some serious chest wounds and we had to stay low. The weather was completely socked in, but these guys needed to get to the hospital. Once we took off, I could not see the ground the whole time. We were on instruments. We used that early instrument system – the one where we got that "Rainbow Card" (AAF Instrument and Approach System SCS 51) When you are in that situation, sometimes it doesn't feel just right, but you have to trust the instruments. I could not see the ground until we were over the runway at Prestwick. When I finally taxied in and shut down the plane, I had sweated more on that flight than I can ever remember.

Ambulances bring the wounded to the airfield during frontline medical evacuation missions.

Right and below: Loading the wounded aboard C-47s during frontline medical evacuation missions.

Right: The caption on the back of this photo reads, *Honey, this is a picture of me and the nurse that used to be on my crew. I wanted you to see her. To look at her, you wouldn't think it, but she is 34 years old.* That probably seemed ancient to 21-year-old Joe Maguire.

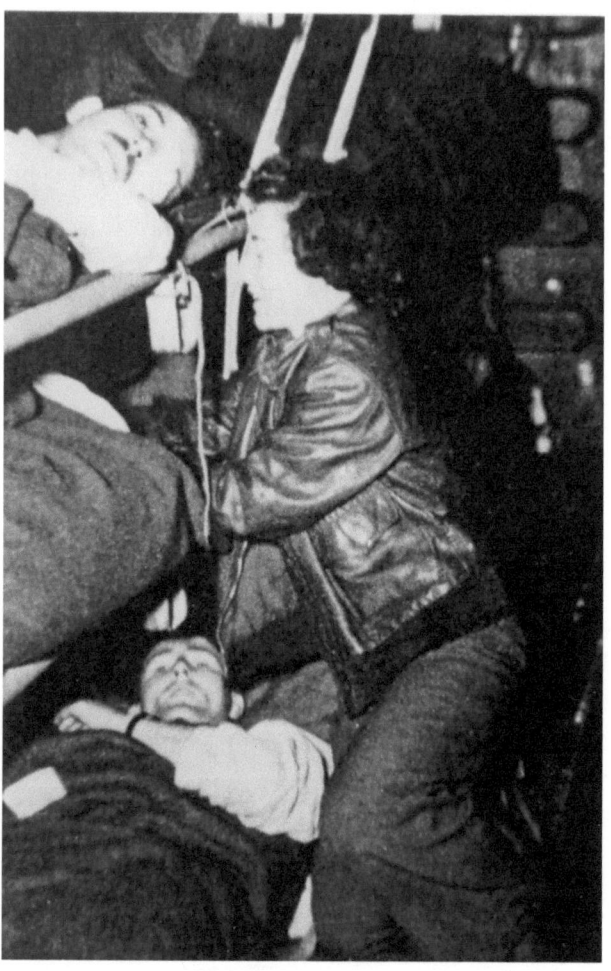

A flight nurse tends to patients on board a C-47.

On February 10, 1945, radio broadcaster Ted Malone covered the 27th ATG and medical evacuation flights on his program, *Westinghouse Presents Top of the Evening*. Here is an excerpt from that broadcast:

> Malone: I have talked to a great number of wounded boys recently and they have all told me that the company aid men, the combat medics who go along with them on every patrol and every advance, are there beside them administering first aid almost as they fall. And minutes later carried on litters if necessary, they start back to the field aid station, the evac hospitals, and eventually the general hospitals in France and England.
>
> A week with the 27th Air Transport Group based here in

France has given me a good chance to see some of the operations that keep this mercy channel of the war functioning.

The first morning, Capt. Vernon Smith of Mobile, Alabama introduced me to the Squadron CO – Maj. James H. Lyons from McAllen, Texas. He arranged for me to fly a supply mission to Belgium to return with wounded. It was snowing briskly and we delayed for a few minutes, during which time I had time to get acquainted with the young major. Snow doesn't bother him. He was flying Army planes to the West Coast three days after the Japs bombed Pearl Harbor in 1941. He was flying P-40 fighter ships to Alaska that winter and was stationed there a while, but I think the assignment that Maj. Lyons is most proud of is a special job hauling wounded boys from the front lines back to hospitals in France and England.

The minute the snow let up, he called the tower for clearance and sent me out to the plane. The ship was loaded with blood plasma and medical supplies needed badly and there were wounded to bring back, so everybody was ready to take off when we scrambled aboard. "Scrambled" is the right word. Try climbing from a snow-covered jeep into an icy plane, with roaring engines blowing your coat off your back – and you are scrambling!

The pilot was Lt. L. H. Sanborn of Merced, California. He was a radio announcer at KYOS back in 1938. He has been in the Army three years, overseas six months. "But that's a long time," Sandy says, "when you have a two-year-old girl home waiting for you." And a lot has happened in that six months, I can add.

One of the longest thirty seconds I ever waited through, ticked off down a runway near Bastogne one day when Sandy's tire blew out as they landed – and to make it worse, for the next couple of days while they waited on a new tire, the Nazis almost captured the airfield.

The co-pilot was Lt. Dick Ramer of Peru, Indiana. He has been in the Army three years, overseas nineteen months, and has had his moments too: Once when he found himself in a formation of Nazi fighter ships coming over to strafe British barrage balloons; and another time when his entire electrical system went out on his ship. He tried to come in blind with no colors of the day (lights)

– no anything. He almost came in by parachute that day, but they made it back. (Note: Aircraft had to display a specific combination of identification lights, which changed frequently, to prevent enemy aircraft from approaching.)

Sgt. Emil Welsh and Fred Beck were engineer and radio operator – and Welsh had a story that amused the rest, but scared me, I'll admit. His plane was lost; the clouds were almost on the ground; his pilot had gone down on the deck to follow a railroad track, assuming it would lead them to some town that they could find on the map. They followed the railroad track alright but suddenly it went into a tunnel in the side of a hill and the pilot had to tip the plane right back up on its tail to keep from flying into the side of the mountain concealed by the clouds. Flying sounds easy and fun – and it is when nothing happens.

The snow was almost a foot deep over the field. The runway had been swept and salted but there was still plenty of ice. We got off all right, however, reached our destination – another field in Belgium – got down safely as the trucks rolled up to unload our supplies. We ate lunch in some captured Nazi barracks. While we were there, a couple of buzz bombs sailed over to hurry us up. But they kept on going out of sight, so we relaxed to finish and returned to our ship. At last, the wounded men went aboard. Two hours later we were back at the field from which we started and our load of wounded was carefully packed in ambulances on the way to hospitals.

Every one of the planes of the 27th Air Transport Group, that brings back wounded, carries a nurse or a medic to take care of the injured aboard the ship. And what stories they will be able to tell when they get home!

Of course, they're usually like Lt. Evangeline Comeau at first, and insist nothing ever happened to them. Lt. Comeau is from Omaha. She has been overseas a year. "Nothing much has happened," she said. Oh at first they slept in tents and sudden rains floated their shoes away, but that wasn't anything. One time a field gave them the wrong heading, or someone wrote it down wrong and they found themselves flying over Germany unarmed. They discovered it when the Nazis started shooting at them, but they got back – so that wasn't anything.

Right after D-Day, when the Nazi fighters were strafing the strips where they flew in for wounded, they had to work pretty fast, but they always got out – work pretty fast unloading supplies and loading on patients.

"How long did it take?"

"Well, everybody helped. Once they emptied all the freight out of the plane and loaded 24 wounded patients on litters and took off in 18 minutes."

"18 minutes?"

"And not a minute too soon, either, because we took off on one end and Nazi fighters started tearing into the other."

"But nothing ever happened huh?"

Lt. Allen W. Plutz is married. She used to live on Long Island. Her husband was in the Army but recently received a medical discharge, so now he's home and she is carrying on the battle. Mrs. Plutz, as the girls affectionately call the blonde young nurse, says the only time she was really scared was when they were taking a load of patients to England and an engine went out over the Channel. Here she was with twenty-four patients in casts and bandages and word came from the pilot to prepare to ditch. That means to prepare to land in the rolling, rocking, choppy English Channel. Of course there are rubber life rafts and life jackets, but the problem of getting twenty-four patients out of a plane and landing them and loading them into rubber life rafts before the plane goes under is really something to think about.

Mrs. Plutz didn't stop to worry. She just began to prepare to ditch. Fortunately, the plane made the English coast with its one remaining motor and they didn't have to ditch. But to say they were worried puts it mildly…

The Chief of air evac nurses attached to the 27th ATG is Capt. Marybelle K. Frazier of Pinole, California. …The only story I could get from her deserves a whole program in itself, but so little of it can be told, I can compress it into a minute.

The only way to get supplies into and wounded out, from some of the garrisons, trapped by counter-attack, is by air. A number of planes that have gone on such operations haven't come back. Some have been shot down and crashed with their supplies and crews bailed out. But one ship carried a nurse from this squadron and that ship never returned. Later, the wreckage was found, burned and twisted and some place in the wreck and ashes someone found a pair of shoes corresponding to the nurse's size. They were her shoes.

Sometime later, in the tent chapel on the edge of the field, the nurses held a memorial service for their comrade who had not come back. Each one told stories she remembered. One mentioned her loyalty, another friendliness and courage. They even told about how neat and careful she was, how wistfully feminine. She would never go to dinner until she had combed her hair just so. One time when the plane had been forced down on a field away from base because of bad weather, she was so put out because she had no bobby pins!

It was the kind of memorial service any girl would be proud to have – simple, unaffected and sincere. They even sang one of her favorite songs to close the service. They thought she would have liked it that way.

Then Capt. Fraser told me the rest of the story. The next morning, a letter came from Reba S. Wittle, prisoner of war in Germany. The nurse! She was a prisoner of war. The shoes had been an extra pair she carried in the ship.

Lt. Wittle is a prisoner of war somewhere in Germany and one line in her note reminiscent of the memorial service I must quote you verbatim. She is the only girl at the prison camp, she confided to the nurses and apparently togged out in overalls or still in her nurse's slacks, she writes: "They finally got me a little lipstick to make me feel a little more feminine."

The stories that Malone captured on his radio program are very typical of WW2 medical evacuation and combat cargo operations.

Around the 21st of February 1945, the 321st Transport Squadron moved to Villacoublay, Paris.

Note: Reba Wittle survived the ordeal and made it home to San Antonio, Texas. Her full story is recorded in *Gooney Birds & Ferry Tales, The 27th Air Transport Group in WW2*, by Jon A. Maguire and The Men of the 27th ATG, Schiffer Publishing, 1998

7. Villacoublay

Young Lt. Maguire flew his first Aircraft Credit Sortie out of Grove, England to A-42 Villacoublay 8 September '44. By late February '45, he was stationed there. The 321st Transport Squadron was billeted in a Chateau in Paris, near the airfield. 3829 Chaville (S.-et-O.) – La Marie E. M. It still stands today.

It was around this time that Joe's cousin Ralph wrote a letter to their old high school buddy, Dolph Farrand, in the CBI (China-Burma-India Theater) 490th Bomb Squadron. Ralph told Dolph about their quarters in Paris and the sights they had visited. Ralph was not in the same place as Joe, but it was similar. Several weeks later Dolph wrote a scathing reply, that basically said, "Oh – You are in a Chateau in Paris? I am in a tent in the middle of Burma…! Dolph had a few choice words for Ralph!

The Chateau at Villacoublay where the 321st Transport Squadron was billeted.

Above: Pilots of the 321st on the balcony of the Chateau. Joe Maguire is standing.

Above right: Lee Powers, Charles Longshore and Bob Bailey of the 321st on the grounds of the Chateau.

More of the 321st: Ray Zych, Dick Ambrose, Charley Longshore and Joe Maguire. The inscription on the back of this one is a bit of a mystery. It apparently refers to a specific period or event and it is written by three of the guys. Perhaps they each had a copy of the photo and inscribed it to each other: "Thanks for helpin' me thru those troubled waters "Chaplin" "Doc" Ambrose. "We'll sweat it out to the end – Chas "Vinegar" Joe Maguire. "Chas" I'll never forget those rugged and ruff times of sweating we went through together. Always remembering "Vinegar" Joe. This was obviously a play on Gen. "Vinegar Joe" Stillwell – perhaps a nickname Joe picked up among some of his 321st buddies.

One thing is very apparent when looking at Lt. Maguire's Aircraft Credit Sortie list: the Allies were closing in on Germany by this point. Every Aircraft Credit Sortie after the move to Villacoublay was flown into Advance Landing Grounds in Germany. The massive effort to supply the needs of tanks and advanced airfields was an overwhelming task. Joe often talked about flying loads of gas, with the airplane reeking of gas fumes. It was a constant source of stress knowing that one bullet could ignite the aircraft. The *Stars and Stripes* newspaper covered their efforts.

Bullet Could Turn Reeking Fumes To Flames

By Arthur W. White
Stars and Stripes Staff Writer

AN ADVANCED SUPPLY FIELD, Germany, April 12. — The workhorses of the Air Force—cargo-carrying C47s—lumbered over the Rhine again yesterday with life-giving gas for the Ninth Army.

Unarmed and unescorted, reeking with gasoline fumes that a single bullet could touch into roaring flames, they flew around the Ruhr pocket to the most forward supply field in northwest Germany.

Before the props had stopped turning the first jerricans were hustled into neat stacks on the field, and within 12 minutes planes were heading back to Belgium for another load.

The C47s on the gasoline run were from Col. Carl R. Feldmann's 27th Air Transport Group. From March 15 to yesterday they hauled a daily average of 54,000 gallons of high-octane gas to front-line airstrips.

In addition, 4,000,000 rounds of .50 caliber ammunition were delivered to tactical air and ground armored units and 1,400 tons of plasma, blood and medical supplies to field hospitals. Forty thousand wounded were flown out on return trips.

Stories about the C-47s carrying much needed fuel appeared in the *Stars and Stripes*.

Sky Tankers Fly Vital Gas to Ninth Army

Jerricans of gas are carried from a 27th Air Transport C47 to P47s, right background, which made emergency landings at the most advanced supply field in northwestern Germany. The fighters, before landing, had just knocked out two Tiger tanks in support of armored columns northeast of Hanover.

The Allies could not advance without gas. The 27th ATG delivered a lot of it! (Credit USAAF)

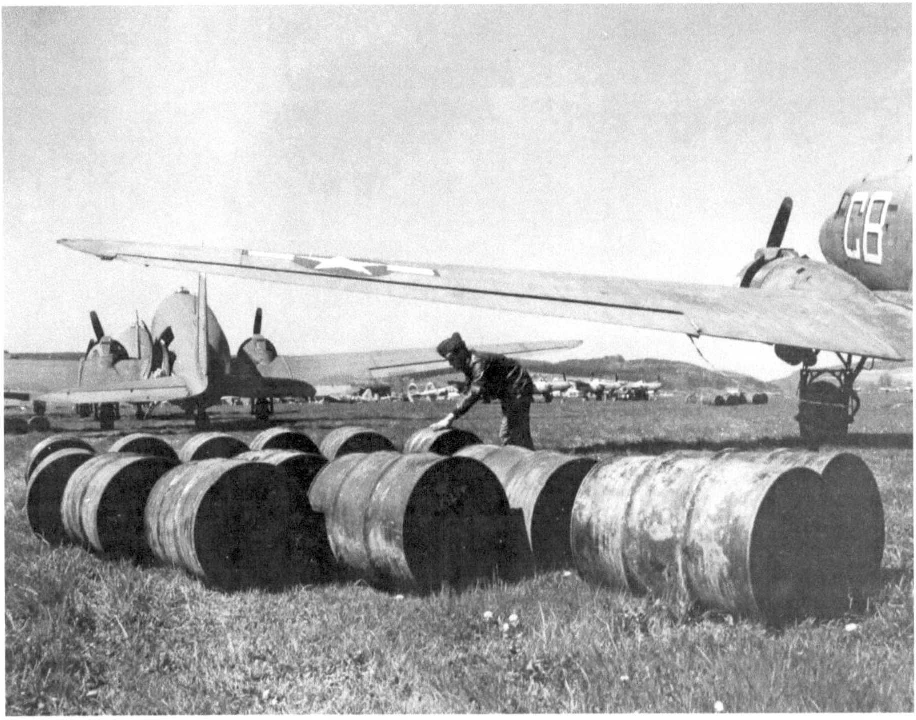
CB

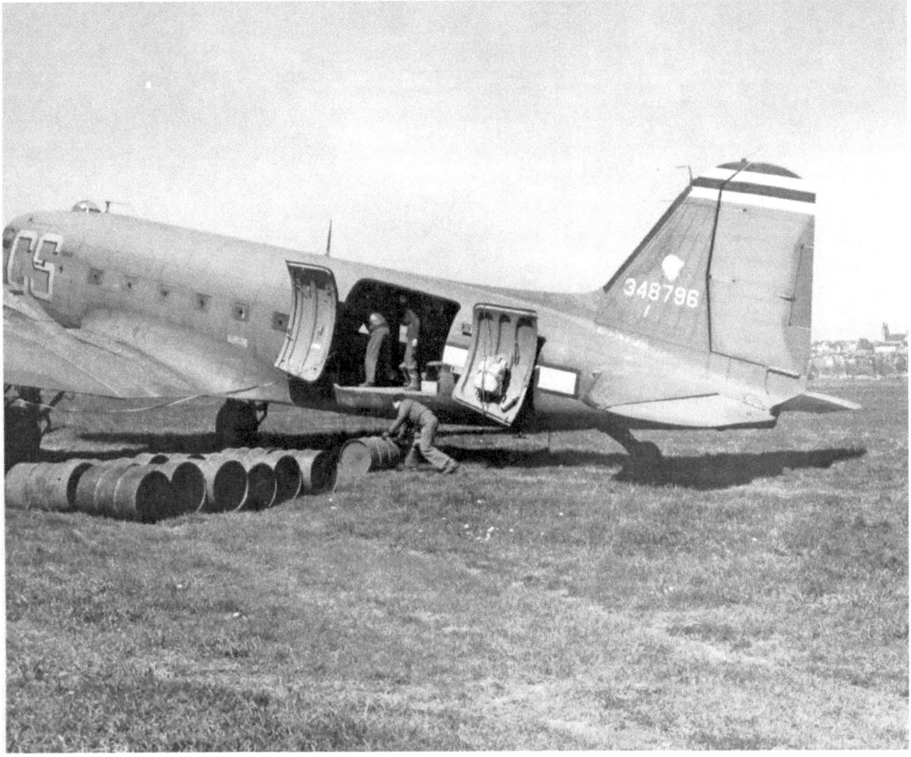
CS
348796

Joe snapped this photo of a US Army Sherman tank heading northeast out of Paris.

The Allies pressed German equipment into service if it was usable. This German tracked vehicle was called a *Kettenkrad*. It was repainted and used at this bombed out airfield at the front.

A US ARMY Jeep being loaded into a 27th ATG C-47.

One thing about a C-47 that was unique from most other aircraft in the inventory was the ability to haul large cargo back to base. Joe recalled:

> *As we had the Germans on the run, a lot of the things they left behind were neat trophies! One trip, we picked up a four-cylinder convertible sports car, yellow and black, with cut down doors. We loaded that in the C-47 just like a jeep. We took that to our motor pool and of course that became registered with the US Government. I guess every military organization under the sun did that, but I think later on they had to try and identify who the rightful owner was. We also picked up a beautiful car that was a Triumph Junior. This car looked like it came off the showroom floor – brand new. It was Pea Green – very pretty. We went through all the labor to get this stuff back, but it was always funny how the commander ended up with it.*
>
> *And then one of the trophies was this motorcycle. It had been an SS motorcycle, black and chrome. I'll never forget it. But it had been in a wreck and the forks were sprung on it* (which he did not know at the time). *I got going on that thing one day and vibration set up in it! You talk about having a tiger by the tail! I didn't think I would ever get the sucker shut down. I couldn't wait to get off of it.*

One of the funniest war story moments I ever witnessed happened 50 years later. Dad was telling that story about the SS motorcycle at our dinner table with my mom and Joe and Colleen Stevens, who were visiting from Idaho. Joe was dad's best friend from the war and they flew on the same crew for a while. As dad was recalling the story, Joe Stevens started to laugh. It was Joe that hit a tree with the bike and then put it back in the motor pool without telling anyone! Stevens' version of the story:

> "One afternoon when I wasn't flying and no one was around to observe, I decided I would try to ride the motorcycle. Though I had never ridden a motorcycle, I knew I could ride a bike – and heck, I could fly an airplane – so what would be the problem? After getting the bike started, I managed to get it in gear and headed off down the road. I hadn't gone more than a couple of hundred yards, when a pole jumped out in front of me, bringing me and the motorcycle to a sudden halt. Picking myself and the bike up, I started it up and headed back up to the billets. I noticed that the motorcycle didn't

want to track straight and after I dismounted, I could see the front forks were badly out of line.

I never mentioned it to anyone until a few years back when I was visiting with my friend, Joe Maguire. He was telling me about the motorcycle and the sprung forks, which nearly did him in when he got up to speed – It was then that I came clean."

I always liked this photo of Stevens enjoying a pint. He looks very tired, yet relaxed. Joe was truly one of the nicest guys you could ever hope to meet and a great pilot. He kept building and flying his own aircraft into his 90s. Dad always said, *"No one could grease a C-47 in on a landing like Joe Stevens."*

In addition to combat cargo, medical evacuation and gas runs, Joe flew a lot of passengers. The 27th was assigned on a number of occasions to fly POWs including, repatriated Allies, German POWs, political prisoners, concentration camp repatriates and VIPs. As the drive across Europe progressed, these flights became more common. Occasionally Joe would talk about these missions, but some of them left very deep scars. You could see it in the way he spoke of them. In later life, he stopped talking about those flights altogether. They were memories he just didn't want to hold on to.

Buchenwald and Belsen (Nazi concentration camps) *were pretty tough stuff that I will never forget…*

I never heard Joe speak in detail of the atrocities he witnessed, but I did hear other men in the group.

Gordon "Gordy" Fisher, who was CO of the 321st Transport Squadron at the end of the war, wrote about his experiences for his local newspaper in Minnesota. There was also an article, "The Inhumanity of Warfare" in *Generations of Today* magazine, January 2010, by Leon Hanson. The 27th ATG flew in reporters to record the horrors at Buchenwald. I am not sure if Joe was on this same trip with Fisher or if he was there on a different mission. Here are a few quotes from Gordy regarding Buchenwald:

> "In late 1944, near the end of the war, Patton liberated Buchenwald, a German concentration camp for political, not military prisoners. This was Patton's first exposure to a concentration camp. He couldn't believe what man could do to his fellow man. The general wanted the whole world to know what took place... We couldn't believe how the bodies were stacked up – like cord wood... Apparently the wife of the commandant was a real fiend, a very sick individual. She had a thing for bright colored tattoos. She told the guards to bring her the tattooed skin of prisoners. She dried and chemically treated the flesh, making shades that she draped over lamps in her home... After seeing the tragedy of war, it became clear why we were fighting. I realized how the German military treated people. Most of those in the camp were Dutch, Danish, French or Polish... It was sickening. Even the survivors were walking around like zombies... It was hard to fathom what kind of monsters these people were."

Gordon Fisher, a man my dad always respected and a great pilot.

Joe continues speaking of flying repatriated POWs:

Each group of military prisoners had somebody that was in command of them who was also a former POW. Most of the time it was the highest-ranking officer from the group. The guy in charge of the group that was going to fly on my plane popped his heels with his heavy British shoes. When you hit those together you can hear them a mile off. The guy was Scottish and had on his kilt. He was very emaciated – skin and bones. He saluted me and offered me a cigarette. That is all he had. I took it. We had been briefed a little bit about what to expect. Any way possible, we were to uphold these people's dignity as much as we could because they needed it. I guess I will be impressed all my life with the dignity that guy had. He basically had nothing, but he was proceeding as if nothing ever happened. As soon as we could talk, he began to talk about his family and that he had been away five years. He told me where he had been taken prisoner and all that kind of stuff. (This man was likely part of the 51st Highland Division, sacrificed at Dunkirk in 1940.)

I will have to say this: I didn't see as bad of things with the military prisoners as I did the political prisoners. They were mostly French people that I saw. You just can't believe that humans can be so inhumane to other human beings. I remember one young French woman who was part of a group of prisoners on my plane. You could tell she had been a very beautiful lady and she was very classy. She was still pretty, but very emaciated. She had a sort of hollow, blank, stare. As we spoke for a while, she shared with me that her biggest fear was that she could never experience emotions again.

I remember the first time we had some German prisoners working for us. They were such fanatics! I was not used to that kind of military courtesy. I would come driving up and they would be there loading the plane with something. Man they would bang their heels and salute! I was so shocked I almost forgot to return the salute! I really didn't care to anyway. At the time I had no respect for them.

I remember one time we picked up a load of German scientists and technicians that had worked on the rocket program. I always wondered if any of those guys went to work at NASA with Wernher von Braun. (This was likely Joe's sortie into R-19 – Nordhausen, 2 May 1945)

A high-ranking German officer POW preparing to board a C-47 of the 27th ATG. Note he has been required to remove the eagle and swastika from his cap.

German POWs awaiting transportation on board a C-47 of the 27th ATG.

German POWs transferring wounded American soldiers from an ambulance to an aircraft.

Commendation from Gen. Eisenhower given to the 302nd Transport Wing, of which the 27th ATG was part, for the role they played in safely repatriating allied personnel.

```
                    Supreme Headquarters
                    ALLIED EXPEDITIONARY FORCE
                    Office of the Supreme Commander


                                                  15th June, 1945.


    Dear Colonel Bateman,

    I wish to commend you and the air and ground crews of
    your Command for the fine job of flying repatriated allied
    personnel which you have done.

    The record of your Wing in helping in the liberation of
    allied repatriates and without a single casualty or serious
    accident, deserves the highest praise.

    Please convey to your staff and to your crews, my sincere
    thanks and heartfelt commendation for a job well done.

                         Sincerely

                                    /s/ DWIGHT D. EISENHOWER



    Colonel Martin A. Bateman,
    The Commanding Officer,
    302nd Transport Wing.
```

Perhaps the most memorable day in Joe's combat flying career occurred near the end of the war, while flying out of Le Bourget to a front-line field at Gütersloh, Germany on April 12, 1945. It was a day that nearly cost him his life. As he grew older, especially in the final year or so, he became almost hyper-focused on that day. He told the story frequently and mentioned it in conversation. It clearly was a day that could not be comprehended unless you were there.

Gütersloh was used as a Defense of The Reich base for NGJ2's Ju 88 night fighters. As a Defense of the Reich base and a place known to be visited by Reichsmarschall Hermann Göring, it was well-defended. The Americans liberated the field in April of 1945 and it was designated Advanced Landing Ground Y-99 Gütersloh. The 363rd Tactical Reconnaissance Group operated from there in late April. The 370th Fighter Group started operations there as well. The first aircraft into the field were C-47s bringing in supplies to prepare for the fighter and recon operations. Joe spent a lot of time thinking through that day as he prepared it for inclusion into the unit history in the mid-1990s. We talked to other members of the 27th ATG to assemble as many facts as possible. Also included are portions of the original combat report…

Our orders were to fly a load of 55-gallon drums of high-octane aviation gas to Y-99, a recently captured Luftwaffe airfield at Gütersloh, Germany, which was now one of our frontline fighter bases. At the time, we young "Supermen" felt these missions were routine and they were flown with little anxiety. In reality they were quite dangerous, as one well-placed round could ignite the whole plane.

We took off, loaded from our field at Villacoublay, Paris, flying to Liege, Belgium, and turning to Gütersloh. We could see several tethered balloons off to our right and soon after, 16 P-47 fighters, in flights of four aircraft each, passed below and in front of us. The air was filled with smoke and the smell of cordite. Thank goodness we did not see any German fighters and vice versa! The Luftwaffe was getting pretty scarce this late in the war.

We located and landed at our destination. Our drums of gasoline were soon unloaded and we were back in the air.

I was first pilot with the load coming in, and Lt. Barker, who was newly transferred to our squadron from another squadron, was co-pilot. One of my crew was Sgt. Bouchier and, regretfully, I cannot remember the other crewman's name. We changed crews about every flight, which makes it difficult to remember a specific flight.

On the trip out, Barker was pilot and I was co-pilot when we were hit with flak. I estimated our position at that time was between five and six miles north of Wessel, Germany. Also we had picked up our heading towards A-93. At that moment, Lt. Barker said, "They are shooting at us from off our left wing!" He turned sharply to the right away from the fire and hit the deck. It was at that moment we were hit in the left wing. Wham! Bang! It was a solid flak hit, which severed the control cables. There was a temporary lapse in conscious thinking while survival instincts took over. I lost our exact location, but was well-aware of our general vicinity.

Not knowing the extent of the damage, Lt. Barker immediately ordered everyone to prepare to leave the ship. The crew and I then put on parachutes, while Lt. Barker trimmed the ship and put it on the autopilot. Then I went to the cockpit to take control while Lt. Barker put on his parachute. I flew the ship for about five minutes, standing between the pilot and co-pilot seats. I decided I could control the ship so I removed my parachute and sat down in the pilot seat.

At the time I took over the ship we were flying a heading of approximately 50 degrees, which I continued to fly for approximately 20 to 30 minutes. All this time I was climbing, knowing there was a possibility of either of the engines having been hit and going out. We did not want to go back through the area where we had been so I started gaining altitude and heading for England. While gradually gaining altitude, we passed over the north side of Arnhem and were over the Zuider Zee on the north edge of Amsterdam.

About that time, Sgt. Bouchier or Lt. Barker yelled, "Here it comes again!" Out of nowhere a fighter hit us with a 20mm cannon shell in the center of the fuselage behind the bulkhead where the radio operator sat. The crew believed it to be an Me 109. The explosion of the 20mm shell left the cargo area looking like a saltshaker. The German made two more passes from behind our airplane. As we would see his tracers, I would cross-control and slide away from the path of fire. The left wing got riddled but the self-sealing tanks worked well. (That was the first time that I knew we had self-sealing tanks!) By now, we were heading out to sea and the fighter broke off the attack. Talk about "luck o' the Irish" no blood was drawn. Fortunately, at this stage of the war, the

Luftwaffe did not have many experienced pilots left. If that fighter jockey would have been half-decent we would have been history.

As we approached England, we got on the World Guard (emergency radio) and hollered, "Mayday Mayday!" I had a conversation with the English operator, in the course of which, he asked, "How many angels are you?" In my excited and frustrated state, I told him there are none on the plane! I, of course, soon figured out that he wanted my altitude. After giving him my best estimate of latitude and longitude, we were soon met by Bristol Beaufighters and an RAF PT type rescue boat. The Beaufighters led us to RAF Woodbridge emergency airfield on the coast of England. I remember at that point, my greatest concern was whether or not my tires were shot out. I buzzed the tower and they told me the tires looked OK, but I decided I better come in hot, just in case. As it settled down, the tires were fine.

We went into the base and met an English interrogator. He said something to the effect of, "Well I guess you lads have had it today. I hate to be the one to tell you, but President Roosevelt died today." On top of that my home city of Oklahoma City was hit by a tornado that day!

50 years later, I learned at a 27th ATG reunion, that the Gooney Bird we were flying that day was used for spare parts.

Flak damage in the wing of the C-47 flown by Joe Maguire and William Barker (aircraft number 3516, squadron/aircraft code CC).

Joe Maguire and William Barker shaking hands over the flak hole in the wing of their aircraft.

Shrapnel holes from the explosion of the 20mm shell are visible in the radio compartment of the ship flown on 12 April 1945.

Several nurses and the guys from the 12 April mission to Gutersloh signed this single note "Short Snorter:" Sgt. E. P. Bouchier "Boo-koo flak." Lt. William B. Barker "We sweated it together." "Big Jim" Milford or Albert Neverlin may have been the other crewmember that day.

Radio operator Bill Champion was not on the mission that day, but he was back at Villacoublay, monitoring the radio chatter: "We had planes shot down on final approach by Me 109s. We had one "happy warrior" take a plane back to England from France with heavy battle damage. They made it by sheer guts and good flying "on the deck." It was hard to put your hand on the wings of that plane without putting it over a flak or bullet hole. I monitored the radio operator's transmission while they were in flight. His "fist" was understandably so shaky it was hard to discern what he was trying to send."

Sgt. Bill Champion. In addition to being a radio operator, Champion also was a Link Trainer Tech.

Joe recalls an occasion when he fought back:

> *I remember another time we were flying in supplies to one of those advanced bases in Germany. At that time of the war, some of those fields were swapping hands pretty quickly. In this case we were given a heading and you had to stay on it as it was supposed to be clear. Well, we were on final approach with a full load. We had the flaps and gear down and we started drawing ground fire! The doggone Krauts had either taken the field back or had not been driven out yet! I sucked up the gear and flaps and poured the coal to that old Gooney! We headed straight across the field. Because of the situation, we decided that the safest route back was probably the way we came in. We were always frustrated because we couldn't really fight back, but we had noticed some German aircraft and equipment on the field. The C-47 was equipped with a long, trailing-wire antenna that you could let out for communications and then it would retract back into the airplane. It had a lead weight on the end of it so it would trail behind the plane. We decided we would let that out and drag it through that line of equipment. We came across that field low and hot and drug that lead weight along that line of equipment. It finally snapped the cable, but whatever we hit, it tore it up! We fired our .45s out the windows. I'm sure it didn't do any good, but we felt better. The crew chief was not happy about that antenna when we got back...*

Joe Maguire's best friend, Joe Stevens, was also hit by flak on a mission with F/O Bob Bailey as co-pilot. Joe Stevens described it:

> "It started as a routine flight, except for the weather, which was overcast skies with ceiling 300 to 400 feet en route and the destination Luxembourg called for ceiling of 1500 feet. The crew consisted of me as pilot, Bob Bailey co-pilot, John McCann crew chief and Alphonse Cogozza radio operator. (McCann was the original crew chief on the crew with Stevens and Maguire with Troop Carrier Command, on the trip across the North Atlantic.) We had a French lieutenant as a passenger, who was with US Intelligence Service.
>
> The flight was made on top of the overcast with planned time and distance letdown after passing a radio fix some miles south

and west of Luxembourg. A letdown was initiated after passing the fix. Descending through 1500 feet we were in the clear. Upon descending down, we could see what looked like the German Siegfried Line of pillboxes, so a climb was started and a return to the radio fix. A new approach was started with a more rapid rate of decent and a shorter time. I was confident that this would put us right over our destination.

Upon descending into the clear, right over an airfield, I was surprised to see flak filling the sky! A climb up through the overcast was immediately made and when we broke out on top, it seemed like we were surrounded by bursts of flak! We were hit and the left engine damaged, rendering the throttle and prop control useless. One burst in the left wing was just outboard of the wing tank. There were several bursts in the fuselage, one hitting the radio compartment, wounding the radio operator in the leg, but not seriously.

Realizing the precarious situation we were in, the French lieutenant began tearing up documents in his briefcase.

After assessing the damage as best we could, we decided to head for a fighter strip near Verdun. When we arrived over the airfield at Verdun, we were advised that we were below minimums. After I explained our situation, we were given permission to land. The ceiling was about 500 feet with one-mile visibility. A letdown was made and we came out right over the airfield. A close in pattern was made to keep the airfield in sight. The landing was uneventful as the landing gear and brakes operated normally, although we had no way of knowing this beforehand.

I can remember wishing I had a manhole cover for a seat cushion to protect the future family. Maybe that wasn't too bad a wish as I still have a large piece of flak that was lodged in my seat cushion as a souvenir of the flight.

As a side note, radio operator, Alphonse Cogozza, was killed about a week later, 18 January 1945 in a C-47 crash near A-54. He was the radio operator on a crew with Lt. Roy "Pork" Shilling, as pilot, which crashed with a load of flight crews who were rotating back to the States. All on board were killed except four passengers."

Radio operator, Sgt. Alphonse Cogozza, and co-pilot, F/O Bob Bailey, standing beneath the flak hole in the left engine cowling that knocked out the throttle and prop controls on their C-47 when flying with Joe Stevens.

One advantage to the days at Le Bourget and Villacoublay was the location of these airfields. There was easy access to the city of Paris. The pilots could go into the city and enjoy time off. On one of these visits, Joe was at Notre Dame Cathedral and a French woman placed a St. Christopher's medal around his neck, which I still have today. It was a beautiful gesture from a French woman who was obviously thankful that the Nazis were gone from her home.

Joe Maguire in Paris (left) standing in front of the *Arc de Triomphe de l'Étoile* and (above) with the Eiffel Tower.

After the April 12, 1945 mission Joe would fly eight more aircraft sorties into Germany and numerous other supply, cargo and transport missions. I heard one of the pilots, who arrived overseas before Joe, talk at a reunion about the nature of the everyday flying. Some of them felt like what they were doing was not important and they wanted to be in the action. This particular pilot was talking about a day when he went to ops and was told to fly to a large supply base and pick up a load. As they were loading the plane, he noticed that it was nothing but cans of black and white paint. He began to complain about the nature of the job and that he wished he could do something important for the war effort. He would later learn that his cargo was used to paint black and white invasion stripes on the allied aircraft that would participate in the D-Day invasion on 6 June 1944. The stripes were painted on the wings and fuselage of nearly every Allied aircraft to help prevent them from being attacked by friendly forces. That black and white paint helped save the lives of Allied airmen participating in the invasion. What they did every day was critical to the Allied victory. According to Gen. Eisenhower… four other pieces of equipment that most senior officers came to regard, as among the most vital to our success in

In these last days of the war, flying across Germany and supporting the advance, there were many interesting aircraft on the fields. Several men of the 27th had their photos taken on the tail of this German He 111 bomber, in snow camouflage. This is Joe on the tail.

This Ju 52 was the German's primary transport aircraft. Joe ran over for a quick photo of the aircraft his enemy counterparts were flying.

Joe was always fond of the B-26 Marauder. His friend Wendell Hoppers flew a tour in B-26s before transferring to the 321st Transport Squadron. Joe took the opportunity for a photo of a B-26 on a frontline airstrip. There is a 27th ATG C-47 in the background. That was likely the ship he flew in with gas or supplies for the B-26 unit.

Africa and Europe were the bulldozer, the jeep, the two-and-a-half-ton truck and the C-47 airplane.

Germany surrendered May 7, 1945. The 27th ATG continued with flights picking up POWs, American soldiers, VIPs and all other activities as the hostilities ceased. The pilots of the 321st Transport Squadron volunteered to go to China, as

Japan had not yet surrendered. Joe was part of that contingent:

> We were at a staging area near Laon, France. We were back living in tents. We had all volunteered to go to China when we got on the train to go to Marseille. At night at a train station in Dijon, we heard a Frenchman selling papers, shouting, "Japan Kaput!" That's how we learned they had surrendered. (10 August 1945) One of our guys was from Louisiana and could speak French. He went and spoke to some of the locals. He came back and said, "Well, we dropped some kind of a big bomb and the Japs quit." That was the first we knew about it.

23 August 1945, the men boarded the USS General C. G. Morton, AP 138, a US Navy Transport ship, and headed for the States. They arrived at Newport News, Virginia, on 2 September 1945.

Crowded conditions are quite apparent on deck of the USS *General Morton*.

Left: Lt. Maguire and Lt. Helen Holycheck, a nurse on Joe's crew, on board the USS *General Morton*. Heading for home!

Below: Pilots of the 321st Transport Squadron on the trip home: Charles Longshore, wearing glasses; Harry Protzeller, center; and Jim Rinehardt, right. The guys in the back, wearing hats, are unknown.

8. Clothing and Equipment

THE AUTHOR HAS WRITTEN several other books about US Army Air Forces uniforms, insignia and equipment. In the course of writing those, I interviewed my dad and consulted with him. I recorded one of our conversations, and thought the reader might find his comments interesting.

Did you ever wear a seat-pack parachute in a C-47?

I don't think I ever wore a parachute after I got out of training. We just had them on the plane. We weren't even sure about which one was ours. (He did mention putting on a parachute briefly, when the plane was hit by flak on 12 April 1945)

What flying kit did you wear in a C-47?

Once we got overseas, we wore a pair of pink or green trousers, shirt. Sometimes we had a dress uniform with us and sometimes we didn't. But we always had those flight coveralls on over whatever uniform we were wearing. We lived in those coveralls. We usually had a jacket. I liked the B-10. I also liked the B-9 or B-11 parka when it was really cold. Sometimes even an Ike Jacket. The scarf (speaking about the white scarf visible in several photographs) *was just to keep my shirt from getting dirty. I would put it around my neck. I think everybody had a piece of a parachute. That's all those scarves are. You gotta remember, some of that clothing may be worn for a year or so before you had a chance to get it cleaned... Those pink trousers didn't have much of a crease left in them. I can tell you those probably had been worn for six or seven months. Where were you going to get them pressed? They wore like iron. I think I got them cleaned once and cleaning consisted of dunking them in 100-octane gas. They smelled like gas a long time and my legs were always red. I guess it didn't hurt the material.*

What about the A-2 Jacket (Summer flight jacket made from leather)?
> *I probably flew with it some. I never did think an A-2 was all that comfortable when it was really cold.*

Did you ever fly on oxygen in the C-47?
> *Most of the time if we did, it was just for fun or if we got way up high. Sometimes if we got a headache we would breathe some oxygen.*

What altitude did you fly most of your missions?
> *Most of the time 5 to 9,000 feet.*

What headgear did you wear?
> *I did not wear a flight helmet very often after primary. Most of the time I would just wear my cap or overseas cap. We were very informal. It was just going to work every day.*

What type of shoe did you prefer?
> *Either the ones I had made for me, or GI shoes. The ones I had made at College Station looked kinda like riding boots but they just covered my ankles. I guess I wore those out. They were even shorter than a Wellington boot. They had a strap and a kiltie that looked like a wide spur strap. I lived in them. They were made from scratch to fit my foot (6 ½ C). (Joe had these made at the same shop that made the boots for the Corps of Cadets at Texas A&M. He was US Army Air Force at the time, but was at A&M as part of the College Training Detachment).*
>
> *We used to go into Burtonwood, which was a big supply base. The big thing we would try to get there was sleeping bags. They were a premium. I called them mummy bags. You crawled in and zipped up. They fit you like a body bag. They were kapok in the interior — very light and then you would put a waterproof cover over them. You could have gotten in the snow in those things. They were a great piece of equipment.*

Was it hard to get equipment?
> *It was tough. If you lost your A-2 it would not be all that easy to get another one. I think everybody guarded their stuff. I don't think anybody would have tried to take anything in my unit. I was small too. There weren't too many people that could wear my stuff. I can remember people*

missing stuff, but I never had any trouble.

Was the A-2 the premium garment?

Well it seems like everybody wanted one. These pictures are pretty self-evident – you can see a lot of us are wearing our B-10s. The 10s were softer, the pockets were easier to use. The B-10 had nice inside pockets. It had a place for pencils. It just had more utility than the A-2. I guess someone constantly weighing the risk of being shot down, that A-2 with a sweater underneath might be a better option. If you had to hide out in rough areas, the A-2 might protect you better.

What about the B-3 (Heavy sheepskin coat)?

I never wore that outside of training in open cockpits and about the only place you ever saw that was on bomber bases.

Did you need a blouse (uniform) to get into the officers club to eat or could you go in flight gear?

It seems like every time I went, I was in proper dress, but I might have had lunch in flight gear. Most of the time when we went into another field, we weren't "lolly-gigging." It's not like the business world. You were getting right on with the program. Your crew had orders to carry out and you had responsibilities. You may land and if you had time you might tell the crew to go to the mess hall and eat. We had a lot of K rations and when we could, we would get our hands on paratrooper rations. That's all we ate a lot of times. Just rations. We might eat in the air. We had a lot of days where we were in the air eight hours or more. You might go over on a two- or three-hour flight and come back and then go on some long hall. When you are flying 90 or more hours a month on average and you are not in the air every day and you have bad weather sometimes so those days are out ... you are flying quite a bit.

Joe was always a "Clothes Horse." He appreciated quality fabric and fine tailoring. Because of his size, he often had to have his clothing altered. When he received his commission, Palace Clothiers from Kansas City came to take orders for clothing. Officers bought their own uniforms. They did receive a clothing allowance. Some chose to buy the cheapest uniform possible and use the rest of the money for other things. Joe bought the very finest quality available. He had bullion wire insignia made for his summer uniform, because he thought the weight of a metal wing would stretch the fabric. He wore his sterling pilot's wing on his olive drab "pinks & greens." The uniforms look just like the day he bought them 75 years later, except the bullion wire is tarnished.

7. Almost Fifty Years Later...

1st Lt. Joe D. Maguire was separated from the United States Army Air Force at Baer Field, Ft. Wayne, Indiana on 9 December 1945. His MOS (Military Occupational Specialty) was Pilot Twin Engine 1051. The battles and campaigns he participated in were: Central Europe, Northern France, Ardennes and Rhineland. At the time of separation he received the European African Middle Eastern Service Medal, the American Theater Service Medal and the Victory Medal.

In 1994 he would receive the Air Medal with 6 oak leaf clusters and the Distinguished Flying Cross, which were awarded for action from 26 August 1944 to 2 September 1945. Many of the men of the 27th Air Transport Group did not receive the medals they deserved during the war. It was through the efforts of Group Historian, Marvin Diehl and General George J. "Jim" Eade, USAF (Retired) that this situation was partially corrected. In General Eade's letter of recommendation to the USAF, he states, "The 27th ATG had a number of detachments that moved around the theater, often with little contact with higher headquarters, which itself was pretty fluid. While not surprising, it is nonetheless regrettable that the administrative process failed to cover all air crew members who deserved timely consideration for the missions they flew." I clearly remember telling dad that I thought it was great that I could be there to see him receive his medals. He said, "Yes, but my mom and dad and my brothers never knew." I know he would have liked to have worn the ribbons on his uniform as a young man. He deserved that. He often wore his DFC lapel ribbon on his navy blazer. Most who saw it, never knew what it was. There were approximately ninety-eight Air Medals awarded to 27th ATG members in 1994. There were only five Distinguished Flying Crosses awarded. Joe Maguire was one of the five, as was his pal, Joe Stevens.

Approximately two weeks before his death on January 16, 2016, he was awarded the *Legion d'Honneur* from France, for his participation in the liberation of France. Joe stayed in the USAF reserve until 1959 when he was honorably discharged as a captain. He did not intend to leave the Air Force Reserve. He was travelling for work at the time and when the papers came to re-up, he placed them on a shelf and forgot about them. He was very disappointed when his discharge came in the mail. He could no longer go to Tinker Field and check out his beloved AT-6 and fly.

General George J. "Jim" Eade, USAF (Retired) who was instrumental in securing long overdue recognition for Joe and others from from the 27th ATG.

Documents for Joe Maguire's Air Medal with 6 oak leaf clusters, and for his Distinguished Flying Cross, both awarded in 1994.

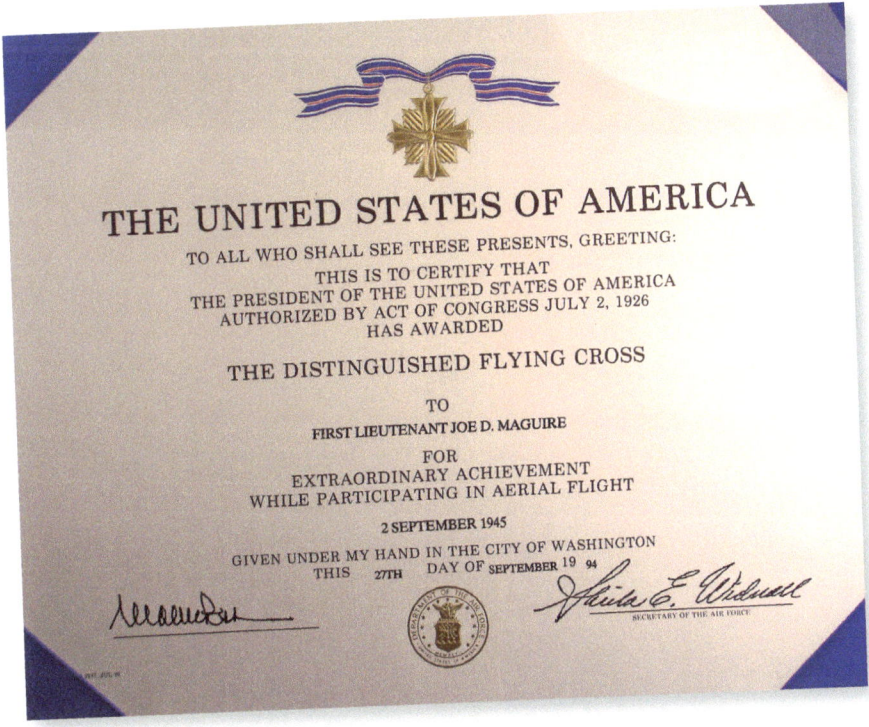

Article from *The Air Force Times* regarding the DFC award.

Dedicated: Joe D. Maguire, left, and Joseph W. Stevens, who received medals belatedly, attend the dedication of a plaque for 27th Air Transport Group at the Air Force Academy in Colorado last August.

Better late than never

This distinguished flier finally gets his medals

By Vago Muradian
Times staff writer

WASHINGTON — Five decades after flying 35 combat missions in Europe during World War II, Capt. Joe D. Maguire finally received eight medals for valor in combat.

Maguire, 71, didn't get his medals in a ceremony but in the mail in mid-October. A package contained his Distinguished Flying Cross and seven Air Medals.

"Needless to say, I was pleased to get them," Maguire said.

He didn't know he was supposed to get the awards until attending a reunion recently.

The awarding was bittersweet.

"My parents are long gone and my brothers are long gone, and I'd have given anything to have them know about it," he said. "But I have children and grandchildren, and I hope that it will stimulate their patriotism when they look at them."

Maguire

Maguire was among four C-47 Skytrain pilots and crewmen with the 27th Air Transport Group of the U.S. Strategic Air Forces in Europe who in October were awarded the Distinguished Flying Cross. The 27th supplied battlefield units with supplies and evacuated wounded soldiers from front-line airfields.

The other three winners are: a retired Air Force general, George Eade of Healdsburg, Calif.; Joseph W. Stevens of Lewiston, Idaho; and Frank H. Simpson of Cornelius, Ore.

Air Force Secretary Sheila E. Widnall approved the medals in September.

"What we're finding is that when World War II and other wars ended, people who left the military left very quickly and didn't realize they were qualified for an award," said Capt. Mike Rein, a spokesman for the Military Personnel Center at Randolph Air Force Base near San Antonio. "All we need is some kind of documentary proof that you were there, did what you said you did and met the qualifications."

Maguire, a native of Oklahoma City, joined the Army Air Forces in 1943, fulfilling a childhood dream of becoming a military pilot.

He earned his silver wings April 15, 1944. But instead of becoming a fighter pilot, he was assigned to fly the C-47 transport.

"When you were assigned to it, you hated it and asked, 'What did I do to deserve this?'" he said. "But when you flew it, you felt really fortunate. ... It was one of the finest airplanes ever made. You could count on it."

The legendary sturdiness of the C-47 saved Maguire's life April 12, 1945. After Maguire's plane dropped off some gasoline for Army Gen. George S. Patton's tanks in Germany, the left wing took an 88mm anti-aircraft round that severed the control cables.

"It flew very sluggishly," he said. "We didn't make any sharp turns trying to get back."

Instead of returning to his home airfield near Paris, Maguire decided to use an emergency landing strip at Great Dunmow, England, in case he had to crash-land.

But the problems weren't over. While crossing the English Channel, a German fighter made three strafing runs against his unarmed plane, scoring two hits.

A 20mm shell from the fighter punched into the plane's radio compartment but miraculously, none of the plane's four-man crew was injured.

Although he left active duty in September 1945, he joined the Air Force Reserve and retired in 1958.

Since the war, he has been back to Europe twice. The starkest memories he has are of the countless white crosses that mark the graves of Americans killed during the war.

"I feel there are so many still there that are deserving of these kinds of awards."

Oklahoma pilot receives top honor from France

BY ADAM KEMP
Staff Writer
akemp@oklahoman.com

For his service as a C-47 pilot in Europe during World War II, Joe Maguire, 92, received one of France's highest honors on Sunday.

Honorary Consul of France Grant Moak presented Maguire, of Oklahoma City, with France's highest distinction, inducting him into the Legion d'honneur (Legion of Honor) for his service in the liberation of France.

Maguire was appointed to the rank of Chevalier (Knight) of the Legion of Honor, which is the highest honor available to those who are not French nationals.

"It was very unexpected," Maguire said. "I had known it existed, but not for me. I thought, 'This is hard to believe.'"

Maguire flew 35 combat missions, accumulating nearly 1,000 flight hours during 12½ months overseas, fighting in Central Europe, the Rhineland, northern France and the Ardennes.

Maguire also has been awarded the Distinguished Flying Cross, the Air Medal with six oak leaf clusters, the European-African-Middle Eastern Theater Service Medal, the American Theater Service Medal and the Victory Ribbon.

Joe Maguire, a 92-year-old Oklahoma City native and World War II veteran, was presented with France's highest distinction Sunday as he was inducted into the Legion d'honneur (Legion of Honor) for his service in the liberation of France.
[PHOTO PROVIDED BY ZACH GRAY, FOR THE OKLAHOMAN]

Article from *The Daily Oklahoman* regarding award of the *Legion d'Honneur* from France.

Left: Jeanan and Joe Maguire following the award ceremony of the *Legion d'Honneur*.
Right: Joe Maguire with Honorary Consul of France, Grant Moak, who made the presentation.

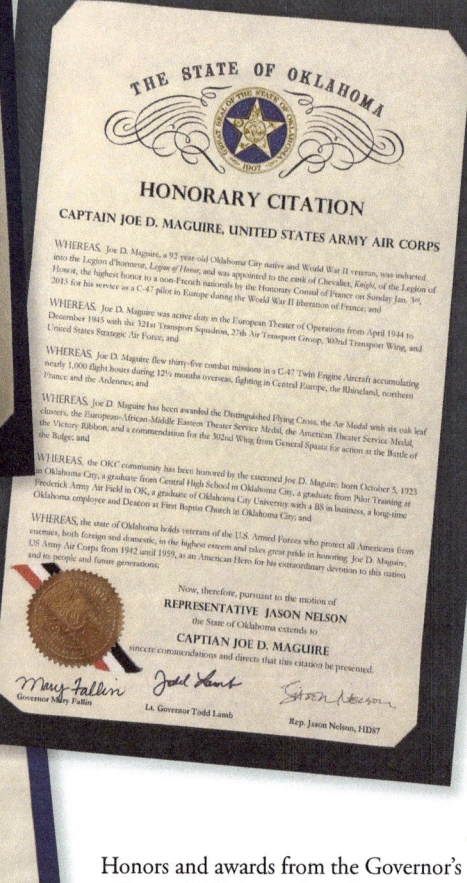

Honors and awards from the Governor's office and Oklahoma State Government recognizing Joe Maguire's service.

Blue skies Dad. We are doing fine but we sure miss you. I'll see you soon.

Appendix 1:
Aircraft Credit Sorties

2nd Lt. Maguire's 35 Aircraft Credit Sorties:

From Grove:

    8 Sep 44 Cargo and Evacuation A-42 Villacoublay France
    10 Sep 44 Cargo and Evacuation A-76 Athis, France
    10 Sep 44 Cargo and Evacuation A-64 St Dizier, France
    13 Sep 44 Cargo and Evacuation A-64 St Dizier, France
    14 Sep 44 Cargo and Evacuation A-64 St Dizier, France
    15 Sep 44 Cargo and Evacuation A-76 Athis, France
    15 Sep 44 Cargo and Evacuation A-55 Melun, France
    17 Sep 44 Cargo and Evacuation A-76 Athis, France

From Le Bourget:

    18 Nov 44 Cargo and Evacuation A-83 Denain/Prouvy, France
    24 Dec 44 Cargo and Evacuation B-58 Brussels/Melsbroek, Belgium
    25 Dec 44 Cargo and Evacuation A-82 Verdun, France
    19 Jan 45 Cargo and Evacuation A-8 Charleroi, Belgium

From Villacoublay:

    30 Mar 45 Cargo and Evacuation Y-64 Ober Olm, Germany
    2 Apr 45 Cargo and Evacuation Y-83 Limburg, Germany
    5 Apr 45 Cargo and Evacuation Y-74 Frankfurt, Germany
    5 Apr 45 Cargo and Evacuation Y-74 Frankfurt, Germany
    5 Apr 45 Cargo and Evacuation Y-78 Bibles, Germany
    6 Apr 45 Cargo and Evacuation Y-79 Mannheim/Sandhofen, Germany
    6 Apr 45 Cargo and Evacuation Y-79 Mannheim/Sandhofen, Germany

6 Apr 45 Cargo and Evacuation Y-79 Mannheim/Sandhofen, Germany
7 Apr 45 Cargo and Evacuation Y-79 Mannheim/Sandhofen, Germany
7 Apr 45 Cargo and Evacuation Y-79 Mannheim/Sandhofen, Germany
8 Apr 45 Cargo and Evacuation R-11 Eschwege, Germany
11 Apr 45 Cargo and Evacuation Y-79 Mannheim/Sandhofen, Germany
11 Apr 45 Cargo and Evacuation Y-79 Mannheim/Sandhofen, Germany
11 Apr 45 Cargo and Evacuation Y-79 Mannheim/Sandhofen, Germany
12 Apr 45 Cargo and Evacuation Y-99 Gütersloh, Germany
18 Apr 45 Cargo and Evacuation Y-90 Giebelstadt, Germany
18 Apr 45 Cargo and Evacuation Y-90 Giebelstadt, Germany
21 Apr 45 Cargo and Evacuation Y-90 Giebelstadt, Germany
21 Apr 45 Cargo and Evacuation Y-90 Giebelstadt, Germany
24 Apr 45 Cargo and Evacuation Y-90 Giebelstadt, Germany
2 May 45 Cargo and Evacuation R-19 Nordhausen, Germany
3 May 45 Cargo and Evacuation R-65 Risstissen, Germany
8 May 45 Cargo and Evacuation R-66 Regensburg, Germany

Original record of Joe Maguire's 35 sorties.

```
                OPERATIONS SECTION
              321ST TRANSPORT SQUADRON (C&M)
              27TH AIR TRANSPORT GROUP
              AAF-180         APO 744

                                            28 MAY, 1945

SUBJECT: AIRCRAFT CREDIT SORTIES FLOWN BY 2ND LT. JOE D. MAGUIRE
         ASN 0702306.

TO    : WHOM IT MAY CONCERN.

         1. IN ACCORDANCE WITH HQ., USSTAF REGULATION 80-6B, PARA-
GRAPHS 1C AND 2L, THE FOLLOWING DATA IS QUOTED FOR YOUR INFOR-
MATION:

       DATE              MISSION                    LOCALITY

      8 SEP 44       CARGO AND EVACUATION            A-42
     10 SEP 44       CARGO AND EVACUATION            A-76
     13 SEP 44       CARGO AND EVACUATION            A-64
     15 SEP 44       CARGO AND EVACUATION            A-76
     15 SEP 44       CARGO AND EVACUATION            A-55
     17 SEP 44       CARGO AND EVACUATION            A-76
     14 SEP 44       CARGO AND EVACUATION            A-64
     10 SEP 44       CARGO AND EVACUATION            A-64
     18 NOV 44       CARGO AND EVACUATION            A-83
     24 DEC 44       CARGO AND EVACUATION            B-58
     25 DEC 44       CARGO AND EVACUATION            A-82
     19 JAN 45       CARGO AND EVACUATION            A-87
     30 MAR 45       CARGO AND EVACUATION            Y-64
      2 APR 45       CARGO AND EVACUATION            Y-83
      5 APR 45       CARGO AND EVACUATION            Y-74
      5 APR 45       CARGO AND EVACUATION            Y-74
      5 APR 45       CARGO AND EVACUATION            Y-78
      6 APR 45       CARGO AND EVACUATION            Y-79
      6 APR 45       CARGO AND EVACUATION            Y-79
      6 APR 45       CARGO AND EVACUATION            Y-79
      7 APR 45       CARGO AND EVACUATION            Y-79
      7 APR 45       CARGO AND EVACUATION            R-11
      8 APR 45       CARGO AND EVACUATION            Y-79
     11 APR 45       CARGO AND EVACUATION            Y-79
     11 APR 45       CARGO AND EVACUATION            Y-79
     11 APR 45       CARGO AND EVACUATION            Y-99
     12 APR 45       CARGO AND EVACUATION            Y-90
     18 APR 45       CARGO AND EVACUATION            Y-90
     18 APR 45       CARGO AND EVACUATION            Y-90
     21 APR 45       CARGO AND EVACUATION            Y-90
     21 APR 45       CARGO AND EVACUATION            Y-90
     24 APR 45       CARGO AND EVACUATION            R-19
      2 MAY 45       CARGO AND EVACUATION            R-65
      3 MAY 45       CARGO AND EVACUATION            R-66
      8 MAY 45       CARGO AND EVACUATION
                     CARGO AND EVACUATION
                     CARGO AND EVACUATION
                     CARGO AND EVACUATION

TOTAL: THIRTY-FIVE (35)    AIRCRAFT CREDIT SORTIES.

         2. THE ABOVE AIRCRAFT CREDIT SORTIES WERE FLOWN BY 2ND LT.
JOE D. MAGUIRE, 0702306,         IN C-47 TYPE AIRCRAFT WHILE SERV-
ING WITH THIS SQUADRON.

         3. THIS PILOT WAS NEITHER WOUNDED NOR INJURED WHILE FLY-
ING ON OPERATIONS IN THIS SQUADRON.

                                        George J. Eade
                                        GEORGE J. EADE
                                        CAPT., AIR CORPS
                                        OPERATIONS OFFICER
```

Appendix 2.
Airfield Locations:

JOE WOULD TALK OF LANDING very near the front lines and in some cases drawing ground fire on final approach or shortly after takeoff. It was not unusual to hear the sound of gunfire or see P-47s looking for ground targets. As I began to research the airfields on this list, it became apparent that he was landing there within days of the field being liberated by the Allies, so it is not surprising that German troops were still in the vicinity. He also said, on occasion the smell of cordite was still in the air. I sometimes thought that he mentioned that for dramatic effect, but after seeing the dates these airfields were liberated, I believe he meant it literally. Flying in a load of 55-gallon drums of gas and drawing ground fire was not for the faint of heart.

Much credit must be given to the IX Engineering Command who was responsible for getting many of these Advanced Landing Ground bases serviceable very quickly. These frontline strips were often made with Marston Mat steel planking. Many of the bases that were seized from the Luftwaffe were heavily damaged due to bombing and shelling. Some had been mined as well. Landing there could be very dangerous for the first planes in. The men of the IX Engineering Command are indeed, unsung heroes!

Airfields

A-42 (Also AAF 180) – Villacoublay is still an active French Air Force base at the time of this writing. It was built before WW2 and was seized by the Germans during the Battle of France in 1940. The Luftwaffe operated from the base until the Americans liberated it 27 August 1944 during the Northern France campaign. Maguire flew his first combat sortie into Villacoublay on 27 September 1944. He would later be stationed there with the 321st Transport Squadron.

A-76 – Athis was a grass airfield constructed by the Germans. It was completed in August 1944. The Americans liberated it in September and it was designated as Advanced Landing Ground A-76. The 36th Fighter Group started flying P-47s out of there in October. It also had a large supply base with access roads. Maguire flew three sorties into Athis in September, just days after it was liberated.

A-64 – Saint-Dizier is a French Air Force base with a long history dating back to 1910 when an aircraft landed nearby. The Germans took over the base in June 1940 during the Battle of France. JG 54, a famous German fighter wing, flew missions from there during the Battle of Britain. The Americans liberated the field in September 1944 and it was designated Advanced Landing Ground A-64. Joe Maguire made three trips into Saint-Dizier in September, shortly after it changed hands from the Germans.

A-55 – Melun-Villaroche was a civilian airport prior to WW2. The Luftwaffe used it after they seized it in June 1940. The Americans liberated the field 1 September 1944 during the Northern France Campaign. Initially the 416th Bomb Group operated from the field until it became a transport and air service depot. Joe flew one sortie into this field in September 1944. Judging by the dates, they were likely combat cargo and fuel runs for the 416th, who were flying A-26s and A-20s at the time.

A-83 – Denain/Prouvy – Valenciennes-Denain Airport is a regional French airport. It was taken over by the Luftwaffe in May 1940 and liberated by the Americans on 11 September 1944 during the Northern France Campaign. Marston Mat was quickly laid and it became operational as Advanced Landing Ground A-83. C-47s started arriving almost immediately, supplying the front lines with combat cargo and flying medical evacuation missions. Joe flew one combat sortie to this location in November 1944.

B-58 – Brussels/Melsbroek, Belgium – When telling the story of the Battle of the Bulge in Chapter 6, Joe mentions that earlier in the day he had flown to Belgium. That trip was to this location. Brussels/

Melsbroek was a field built originally by the Germans. It was liberated in 1944 and operated by the RAF. It was designated Advanced Landing Ground B-58. During Operation Bodenplatte, on 1 January 1945, many allied aircraft were destroyed on this field by the Luftwaffe aerial attack.

A-82 – Verdun - Base Lt. Étienne Mantoux, formerly Étain-Rouvres Air Base is now a French Army Light Aviation Base. This location was one of the most notable missions Joe Maguire and the 27th ATG would fly during the war. It was one of the stories he told most often – Christmas Eve 1944, the Battle of the Bulge (the story appears in Chapter 6). It was built in 1937 and was used by the French and later the Germans until the US Third Army liberated it in September 1944. The US designated the field Advanced Landing Ground A-82. On 13 September 1944, the 7th Field Hospital was located here and it was a hub for evacuating wounded. In October, the field was upgraded and it became a combat cargo supply area with access roads. On 5 November, the 362nd Fighter Group was stationed there with P-47s. They would fly ground support during the Bulge. There was controversy having a combat unit on the same field as a hospital. The hospital was relocated 15 January 1945. Joe flew a sortie into Verdun on Christmas Day 1944 in support of the Bulge. The 302nd Transport Wing received a Commendation from Gen. Spaatz for their action at the Bulge.

A-87 – Charleroi, Belgium – Brussels South Charleroi dates back to 1919. 14 September 1944 it was designated Advance Landing Ground A-87. Maguire flew one sortie to this location in January 1945.

Y-64 – Ober-Olm started as Fliegerhorst Ober-Olm, a Luftwaffe base in 1939. After the Battle of France, it became a "Defense of The Reich" airfield. Prisoners of the SS concentration camp Hinzert maintained it. XII Corps, 90th Division on 22 March 1945, captured the field. Combat Engineers from the IX Engineer Command were able to get it operational by 27 March and it was designated Advanced Landing Ground Y-64 Ober-Olm. In early April the 354th Fighter Group began operations with P-47s. It also became a base for C-47s to move in combat cargo and evacuate wounded. Joe made one sortie here on

30 March 1945, three days after it became operational.

Y-83 – Limburg airfield was a grass airfield, built by the Luftwaffe to protect the railroad yards in 1944. It was captured by US Army forces on 25 March 1945 and designated Advanced Landing Ground Y-83 Limburg. Lowell Thomas of NBC made a broadcast from Limburg. The 67th Tactical Reconnaissance Group operated from there for a short period before it became a combat cargo and evacuation base for C-47 operations. Maguire flew one sortie here on 2 April 1945.

Y-74 – Frankfurt (Fliegerhorst Eschborn) was built by the Luftwaffe in 1940 as a fighter base and was used in the Defense of the Reich. The US Third Army moved into Frankfurt in March 1945. It was designated Advanced Landing Ground Y-74 Frankfurt/Eschborn and became operational as a C-47 combat cargo and medical evacuation base. In early April, the 371st and 367th Fighter Groups began P-47 operations from the field. Joe made two trips to Frankfurt, both on 5 April 1945 and then made a third trip that day to Biblis!

Y-78 – Biblis was built by the 850th Engineer Battalion, very quickly, in early April 1945 on the site of a Wehrmacht Army barracks. It was very important to the Allied forces during the Ruhr Pocket. The 12th Air Force, 27th Fighter Bomber Group operated A-36 Apache ground attack aircraft from the base. The 27th gave up their A-36s in April 1944, flew P-40s for about a month and then got P-47s. 300,000 German troops were captured at the Ruhr Pocket. Joe Maguire's third sortie of the day on 5 April 1945 was into Y-78.

Y-79 – Mannheim/Sandhofen is now Coleman Army Airfield. It is named after Lt. Col. Wilson D. Coleman who was killed in action in France on 30 July 1944. There was a commercial field established in 1925 as Flughafen Mannheim-Heidelberg-Ludwigshafen. In 1926 it became known as Mannheim-Neuostheim. In 1937 the Luftwaffe rebuilt it as Fliegerhorst-Kaserne. One of the more famous German Fighter units operated from this field, JG 53 *Pik-As* (Ace of Spades). In September 1944 a POW camp was established on the site with SS guards. Lt. Maguire made five trips to this location over a two-day period 6-7 April 1945. Joe made three more runs on 11 April to this location.

R-11 – Eschwege was built in 1937 as a Luftwaffe transport base with Ju 52s operating out of it. It continued in use as a transport base throughout WW2. It was captured in April 1945 and designated Advanced Landing Ground R-11 Eschwege. The 67th Tactical Recon Group moved in 10 April 1945. Joe Maguire flew one combat sortie to this location on 8 April. Judging by the dates, this was likely a gas/combat cargo run in preparation for the arrival of the 67th, which would arrive 2 days later.

Y-99 – Gütersloh was forever etched in dad's mind, as he really should never have survived that day (detailed account is included in Chapter 7). Gütersloh was built before the war as a Luftwaffe fighter base. It had a room in the tower of the Officers' Mess known as the Göring (Reichsmarschall Hermann Göring) room. Legend has it that Hermann himself would sit and tell of his exploits and then exclaim, "If I should lie, may the beam above my head crack!" One of the men engineered a beam by sawing through it and then with a system of pulleys, he could make it appear that the beam did crack in response to the Reichsmarschall's stories. The trick could supposedly be performed, even after the British took over the field after the war. In 1944-45 it was used as a Defense of The Reich base for NGJ2's Ju 88 night fighters. The Americans liberated the field in April of 1945 and it was designated Advanced Landing Ground Y-99 Gütersloh. The 363rd Tactical Reconnaissance Group operated from there in late April. The 370th Fighter Group started operations there as well. Post war, the base became RAF Gütersloh. Joe was flying in a load of 55-gallon drums of high-octane aviation gas on 12 April 1945.

Y-90 – Fliegerhorst Giebelstadt was one of the first Luftwaffe fields established in 1935, about 250 miles south of Berlin. It was part of the Defense of the Reich system. Me 262s and Me 163s operated from this field. The American 12th Armored Division liberated the field in late March 1945. The field became operational for the US as Advanced Landing Ground Y-90 Giebelstadt and was immediately used as a C-47 combat cargo and Medical evacuation field. On 20 April 1945 the 50th Fighter Group and the 417th Night Fighter Squadron began operations there. Joe made five trips to this field from 18 to 24 April 1945.

R-19 – Nordhausen was a field near Mittelbau-Dora, which was a forced labor facility by concentration camp inmates. It was a sub-camp of Buchenwald. Dora was founded in 1943 and became the facility for V-2 rocket production. Many of Germany's scientists were at Nordhausen, including Wernher von Braun. Von Braun and his staff surrendered to the Americans and later were instrumental in the US space program. The Americans flew many prisoners out of this camp, as well as V-2 rocket parts. Joe made one trip to this site. It is now Flugplatz Nordhausen.

R-65 – Risstissen was in operation from 27 April to 15 June 1945 as a supply and evacuation field Advanced Landing Ground R-65. Maguire made one trip to this field.

R-66 – Regensburg was open from 28 April to 15 June 1945 and designated Advanced Landing Ground R-66. Lt. Maguire's final aircraft sortie was to this location 8 May 1945. This was not his final flight in theater.

Photo: Zach Gray

Acknowledgments

I have to start by thanking my mom and dad. I was lucky to have great parents who were always supportive and did the best they could for my sister Jana and me. This work would never have been possible without the men of the 27th Air Transport Group who always took time for me, answered my questions and shared their photographs and stories with me. I especially need to recognize Joe Stevens, Gen. George J. "Jim" Eade, Richard Kirkes, John Schneider, Gordon Fisher, Bill Champion, Ralph Jenks, Art Kane, Dick Seebers, Joe Cioffi, Editor of Ferry Tales and so many others. Thanks to Von Gilbert, Westinghouse Public Relations.

The majority of the photographs are from my dad's album, but I also must thank the USAAF, John Campbell of Campbell Archives, Jim Nawrocki, Richard Chancellor and Zach Gray.

Thanks to Sean Maguire, Jon Bernstein, Jeanan Maguire and Jana McNeill, for help with proofing, editing and opinions.

Finally, thanks to my dear friends, Mick and Diane Prodger for always supporting me and listening to me.

On a personal note I'd like to thank all my friends and family, as well as others who had read my previous books and expressed interest in learning more about my dad, for encouraging me to gather up his stories in the form of a biography. Thank you all, especially my sister, Jana, and her husband, Steve; to my wife, Rhonda, and to our children and their families: Kelly, Matt, Maddy and Owen; Kathleen, Ryan, Ely, Evelyn and Ezra; Sean and Steph; Megan, Chris, Tyser, Karslyn and Brynox, I hope this will help you to remember Papa Joe and Nanny Jean, as well as the men and women of the Greatest Generation.

The Author

Jon A. Maguire has been a history buff and collector of Militaria since he was a small child in the '60s. Jon has written a number of books on clothing, insignia and equipment of the US Army Air Forces, as well as the unit history of the 27th Air Transport Group, in which his father served as a pilot during WW2.

Jon A. Maguire and his son, Sean, visiting a C-47 in Wichita, Kansas. This aircraft is painted in the markings of a C-47 of the 321st Transport Squadron, 27th Air Transport Group. The aircraft was assigned to the 27th ATG in September of 1944. There is a real possibility that Joe Maguire had time in this exact aircraft. Betsy's Biscuit Bomber is owned and operated by the Gooney Bird Group, LLC, and maintained by a team of dedicated volunteers. Visit them at the Estrella Warbird Museum located on the Paso Robles Municipal Airport. They had stopped in Wichita on their way home from Daks Over Normandy for the 75th Anniversary of D-Day.

By the Same Author:

American Flight Jackets, Airmen and Aircraft (with John P. Conway)

Silver Wings, Pinks & Greens

Gear Up!

Art of The Flight Jacket (with John P. Conway)

More Silver Wings, Pinks & Greens

Gooney Birds & Ferry Tales, The 27th Air Transport Group in WW2

German Headgear of WW2, Vols 1 and 2 (with Pat Moran)

Uniforms of the Third Reich (with Arthur Hayes)

Silver Wings & Leather Jackets

www.ingramcontent.com/pod-product-compliance
Lightning Source LLC
Chambersburg PA
CBHW040200100526
44591CB00001B/4